H. G. Gierloff-Emden

Orbital Remote Sensing of Coastal and Offshore Environments

A Manual of Interpretation

Walter de Gruyter · Berlin · New York 1977

Author
H. G. Gierloff-Emden, Professor Dr. rer. nat.
Chairman, Institut für Geographie der
Ludwig-Maximilians-Universität München
Abteilung Luftbild- und Satelliteninterpretation
Luisenstraße 37
8000 München 2

Scientific design – H. G. Gierloff-Emden
Graphic arts and technique – G. Edelmann, München

Layout
H. G. Gierloff-Emden and Lilo Gierloff-Emden

Lithography
Colour – SÜSS-DRUCK, Gebr. Pech OHG, Moosburg
Black and white – Ilmgaudruckerei

Library of Congress Cataloging in Publication Data

Gierloff-Emden, Hans-Günter
Orbital remote sensing of coastal and offshore environments.
Bibliography: p. 1-176
1. Coasts-Remote sensing. 2. Oceanography-Remote sensing. I. Title.
GB 452.2.G 53 551.4'6'0028 77-5869
ISBN 3-11-007278-5

CIP-Kurztitelaufnahme der Deutschen Bibliothek

Gierloff-Emden, Hans-Günter
Orbital remote sensing of coastal and offshore environment: a manual of interpretation. – 1. Aufl. –
Berlin, New York: de Gruyter, 1977.
ISBN 3-11-007278-5

 Typesetting, printing and binding: Ilmgaudruckerei, Pfaffenhofen-Ilm.
Printed in Germany.

Dedicated to

the late Prof. Dr. Carl O. Sauer
Department of Geography
Berkeley, University of California

in memory of his fundamental contributions
to the understanding of our world
in the sense of "man and environment"

and to

the late Prof. Dr. Richard J. Russell
Department of Geography and Anthropology
and Coastal Studies Institute
Baton Rouge, Louisiana State University

in memory of his work on coastal environment
and coastal resources

Cover photography: NASA ERTS-1-MSS imagery colour composite from multispectral scanner images bands 4, 5, 7, enlarged section of NASA ERTS-1-MSS 15/1088 ID 1209-15444

Jiquilisco Lagoon of El Salvador

(Vegetation comes out in red colour)
(Colour composite made by Department of Geology, University of Munich)

This manual has been prepared as a guide for scientists in the interpretation of coastal and oceanographic features appearing on satellite photographs and remote sensed imagery. It has been designed to provide assistance for the study, understanding and preservation of the coastal environments.

It is a pleasure to acknowledge the constructive cooperation with the secretary's office of the Intergovernmental Oceanographic Commission (IOC) in Paris during the time of preparation of this book.

München, 1977 H. G. Gierloff-Emden

Errata

The author regrets that the text of this book contains some errata, such as:

Page 19 Landsat 1 made in two years 15 000 orbits (more than 10 000)
Page 20 NASA's Landsat-2 satellite launched in 1972 (1975)
Page 21 Skylab 3 launched in 1975 (1973)
Page 37 signal which is intensified (amplified)
Page 37 the images are sent to earth (transmitted)
Page 37 the successive orbit lies a bit more westerly (lies more westerly)
Page 69 Athmospheric (Atmospheric)
Page 151 Lagoon of Jiquillisco (Jiquilisco)

Contents

(The table of contents has been typographically simplified)

Part I: Orbital remote sensing and coastal studies: general remarks

Introduction

In 1957, nearly 20 years ago, the first man-made satellite was launched, followed five years later by the first manned spacecraft. In 1975 the third Skylab mission took place, the most extended manned spacecraft mission ever carried out; this mission was a great advance in modern oceanography. It is only 100 years ago that modern oceanography began with the research voyage of HMS "Challenger" which returned in 1875 after a three-year oceanographic expedition around the world. The "Challenger" expedition was sponsored by the Royal Navy and the Royal Society of London, and sailed in 1872 from Portsmouth for a three years voyage which was to cover 68 000 nautical miles. A hundred years later, in 1972, the first Earth Resources Technical Satellite – ERTS 1 – was launched into orbit from NASA to serve as a spaceborne platform equipped with orbital remote sensing systems, for continuous earth observation from space. The ERTS 1 with its high spatial resolution imagery has already provided evidence of the value of orbital remote sensing for oceanography.

In the last few years new perspectives in oceanography have been gained, not only by surveillance of the oceans using orbital imagery, but also by visual examination of the bottom of the oceans using photography from manned and unmanned submersibles.

The first books on orbital remote sensing for oceanography were published by: P. C. Badgley, L. Miloy and L. Childs (eds.), 1969, Oceans from space: Gulf Publishing Company, Houston, Texas. And: G. C. Ewing (ed.), 1965, Oceanography from space: The Woods Hole Oceanographic Institution, Woods Hole, Mass.

Orbital remote sensing with satellite photography and imagery can be called the "Third Discovery of the Earth" as the author H. G. Gierloff-Emden wrote in his book: Weltraumbilder, die dritte Entdeckung der Erde, 1974: List Verlag, München (together with J. Bodechtel).

We must remember all the pioneers of orbital remote sensing, including the technicians, and especially all the astronauts.

The material used in this book for the scientific work, presents data and satellite photographs and imagery from the NASA (National Aeronautics and Space Administration), USA, which were gained for the NASA Earth Resources Survey Program.

Coastal resources

"In recent years numerous workshops, conferences, and symposia have been convened for the purposes of identifying, analyzing, and solving coastal zone problems. This current interest in "things coastal" has grown out of an almost endless list of conditions and circumstances. A few of these stimuli are: population pressure, increasing conflict between coastal users, over-exploitation, oil spills, the environmental crusades of the early 1970s, governmental regulations, recreational potentials, the creation of the Sea Grant Program, and the realization that the coastal resource base is in jeopardy.
The coastal zone – the juncture between land and sea – is essentially linear and thus of finite areal extent. Although distinct in concept, it is difficult to define precisely because of the great variability that exists along and across it. This variability comes from contrasting characteristics associated with both the land and the sea, such as geologic structure, climate, biota, water chemistry, waves, currents, and tides. All such elements have a bearing on the nature, distribution, and availability of resources along the coast."

H. J. Walker, 1975 in the outstanding scientific work: "Coastal Resources", vol. XII, Geoscience and Man, pp. VII. Baton Rouge, School of Geoscience, Louisiana State University.
The paper cited demonstrates the great variety of research that can be applied to coastal resource topics, research that will bolster ongoing efforts to utilize the coastal zone and its resources wisely.

The importance of coastal studies is evident as presented through international activities in science

"During the December (1974) meeting, three ad hoc advisory panels were set up by SCOR to deal with specific problems pertaining to the coastal zone:
– Mangrove Ecology and Productivity
– Lagoons
– Biogeochemistry of Estuarine Sediments
The Scientific Committee on Oceanic Research (SCOR) of the International Council of Scientific Unions is an international non-governmental organization composed of representatives from the scientific community who work together to review, and propose solutions to, problems related to oceanographic research. As such, SCOR is also one of the scientific advisory bodies to the Unesco programme in marine science, as well as to the Intergovernmental Oceanographic Commission (IOC). The subjects covered are as wide as is the field of marine science, ranging from fundamental research to the oceanographic environment and methodological problems."

Reprinted from the International Marine Science Newsletter, Dec. 1974.

In the programme of the "Joint Oceanographic Assembly", to be held in Edinburgh, in September 1976, there are several topics which concern the lagoons, estuaries and bays of the oceans, as for instance:
· Natural variations in the marine environment
· Regional studies of dynamics and productivity
· Man and the sea
· Dynamics of ecosystems
Unesco organises regional training courses in *ecology and environmental management* and post-graduate courses for the *study and management of the natural environment.*
Orbital remote sensing with satellite photography and imagery is an important tool in the development of coastal studies.

Remote sensing of natural resources

The role of Unesco's resources research programme

One of the results of the recent development in space technology is the use of space techniques for terrestrial studies, in particular the use of remote sensing from satellites for studies and surveys of the natural resources of the earth.

It is with great interest that Unesco follows developments in the application of remote sensing to the study of earth resources. In relation with integrated studies of natural resources and with the various disciplines of the environmental sciences, Unesco has for a long time been involved in activities linked with the utilization and development of remote-sensing techniques.

Remote sensing from satellites constitutes a new tool of environmental and natural resources research which can be extremely valuable when necessary to survey or monitor large areas of the earth's surface on a relatively small scale or to obtain a synoptic view from space as, for instance, in the case of satellite imagery for meteorological purposes.

Aerial photographs and other remote-sensing data are used in a number of field studies, implemented by Unesco in the framework of UNDP projects as well as under Unesco's Regular Programme. Unesco has an interest in the development of use of satellite data for quite a number of purposes, including such long-term intergovernmental programmes as the Man and the Biosphere Programme, the International Hydrological Decade, the International Geological Correlation Programme, and research and monitoring programmes co-ordinated by the Intergovernmental Oceanographic Commission, such as the International Decade of Ocean Exploration (IDOE), the Tsunami Warning System in the Pacific (ITSU) and the Integrated Global Ocean Station System (IGOSS): (a) for the small-scale mapping of soils and vegetation types (for instance, on the continental scale); (b) for hydrological studies such as the study of drainage patterns, soil moisture, groundwater discharge into the seas, waste disposal, snow and ice cover, etc.; (c) for mapping and studies of geological structures, geomorphological features and geophysical properties of the earth's crust; (d) for measurement of ocean temperature, turbidity, roughness, pollution; (e) for tsunami warning purposes (using geostationary satellites).

Nature and Resources Vol. X., No. 1, January–March 1974

Orbital remote sensing

The significance of satellite photography and imagery for the study of features of coastal zones, as a field for oceanographic research

Among the various sciences which can gain from the analysis and interpretation of orbital remote sensing, both satellite photography and imagery, is that of oceanography in particular in the coastal zones.

"Surface and subsurface features of the oceans can often be seen more clearly from a height than from the surface. This fact, combined with the remoteness of much of the ocean and the high cost of shipboard operation and data collecting – typically $ 4000 per day – makes spacecraft observation economically practical." K. R. Stehling, 1969.

The coastal zones as a field for oceanographic research

In oceanography the fields of research are divided into three basic regions:
1) the upper boundary layer – the ocean – atmosphere interface,
2) the lower boundary layer – the ocean – seafloor interface and
3) the water body of the ocean itself,

but one should add the fourth region – the margin of the ocean as the triple interface sea – land and air or atmosphere, hydrosphere, and lithosphere, i. e. the coastal environment including the marine belt which is in relationship to and dynamic interaction with this continental coastal belt.

Due to the specific character of coastal landscapes, they are extremly flat and extended areas where most static features (tidal flats), as well as the dynamic processes (tidal stream transporting turbid waters), are exposed to the view from above.

Environmental changes generally take place over large horizontal distances in this type of landscape. Zones of relatively high reflectance contrasts (barrier islands, tidal channels etc.) are usually large enough to be discernible on orbital photographs and images (excluding small features, i. e. less than 50 m). The uniform flatness of the topography of the environmental characteristics of lagoons, estuaries and bays eliminates variations in reflectance due to sloping surface and shadows.

"Within easy sight are the waves and swells, biological growth and currents, floating objects and pollutants which occur in the top few centimeters of the water. In very clear water on a sunny day, features can be discerned below the surface to depths of 50-100 ft, often more clearly from some height."

K. R. Stehling, 1969; Remote sensing of the Oceans.
From Astronautics and Aeronautics, May, pp. 62–67.

The fields of application of orbital remote sensing include coastal morphology, coastal boundary surveys, nearshore bottom topography, water composition and depth analysis, tidal wetlands and tidal flats, tidal deltas, tidal currents, marine currents, fish resources and pollution, i. e. both static features and dynamic features (change detection). On these topics there exists already a considerable body of literature.
Remote sensing cuts cross several established disciplines.

Development of orbital remote sensing

This publication presents examples from orbital remote sensing, both satellite photographs taken from manned spacecraft NASA missions Gemini, Apollo, Skylab and imagery from unmanned satellites such as NASA's ERTS 1 (now renamed Landsat). The former are photographs from cameras recorded on films with familiar geometry whereas the imagery from this unmanned satellite is a product of scanners with a special, system-inherent geometry.

Orbital photography and imagery has stimulated our thinking in earth science. These pictures represent, for the first time in the history of man, a synoptic overview of such large areas of the surface of our earth and provide for the thematic mapping of meso and large scale phenomena of environmental features.

A typical vertical orbital photography taken at an altitude of 150 nautical miles will cover approximately 10 000 square nautical miles and have a scale of 1 : 750 000. One scene of ERTS-1 covers 34 000 square km. Such a synoptic overview, covering 100 nautical miles square or greater was, before orbital remote sensing, only available through composites or mosaics of aerial photographs, and not from synchronous photography.

Manned spacecraft orbital photography has some shortcomings, depending on several system-inherent conditions, mostly due to the short length of the missions (time of the year). Hand held cameras (as used in missions Gemini and Apollo) provide imagery of photographic targets of opportunity depending on illumination at the particular moment the photographs were taken (day time); there is little possibility of exact repetition of images in following orbits although several scenes were photographed as proposed targets of repetition on different missions; the possibility of taking a photograph of a particular scene of the earth's surface is restricted due to mission orbits (orbital limitation for Gemini and Apollo was 33°latitude); varying flight altitudes due to elliptical orbits of some missions caused different sensor fields of view and different scales of the photographs. Some orbital photographs taken with hand held cameras are not vertical, but oblique. Skylab mission had constant nominal flight altitude due to non elliptical orbits and delivered vertical photographs. Moreover, manned spacecraft experiments for earth observation are 1000 times more expensive than unmanned satellite experiments. Gemini and Apollo missions with their earth resources programmes did not produce very many scenes from coastal areas, but those there are show clearly coastal features and patterns and coastal and marine processes.

Nevertheless these pictures are valuable documents and represent a wealth of material with considerable potential for interpretation and mapping both for earth scientists and for practical purposes. They are important for use in the scientific development of earth surface- and environmental-phenomena and processes, as used in this publication for oceanographic and coastal environments.

The pictures herein provide some examples of the informative potential of satellite pictures for oceanography and coastal environments.

Pictures taken with hand held cameras also provide imagery to supplement data from other sensors. Development of remote sensing techniques led to multispectral photography from orbit, carried out with the S-0-65 experiments of the Apollo-7 mission. An example of the application of this technique to the coastal environment is presented in this publication.

"Multispectral space photography will be the main optical tool for gathering information in the visible and near infrared region. It is based on photo records taken in several discrete spectral regions, followed by various processes for comparing or combining these records to discriminate or enhance a particular subject, or its main features, by the differences between its spectral reflectances and those of the background."
D. S. Ross, 1969, "Colour enhancement for ocean cartography" p. 50 ff.: in "Oceans from space".

Due to their unique and static character, pictures taken from satellites are valuable for mapping quasi-static features (such as coastlines) and also provide moreover a permanent record of changing processes at one moment of time; they also assist indirectly the study of the dynamics of the processes (such as turbid plumes of water masses), at least by recording one stage of a phenomenon or of a process (such as coastal water dynamics). In terms of remote sensing this is called *"change detection"*.

Change detection. Since the pictorial information value for coastal dynamics depends strongly on one very important moment in the life of the process, which may last for only a short time scale i. e. 5 to 30 minutes (as for instance boundaries of suspended sediment plumes at the beginning or end of tidal stream flows), not even ERTS images with 18 days repetition rate can record this occurrence, as some hand held manned spacecraft orbital photographs could and did, in some cases as classical ones.
(Examples: Gemini and Apollo photographs).
The orbital remote sensing of NASA's Skylab (three missions) provides data taken during three different seasons – June 1973, August and September 1973, and November 1973 through January 1974. With other remote sensing products, it delivered data from EREP (Earth Resources Experiment Package) – orbital photography (ETC camera) and multispectral photography with standard scale precision and exact vertical photography. The high image and ground resolution (also called space resolution) of the ETC (Earth Terrain Camera) surpasses all other orbital remote sensing products. These experiments covered paths of orbits limited to 51° latitude north and south.
An immediate *benefit* of synoptic orbital surveys can often result from the *re-evaluation* of scientific interpretations which require re-examination because of the local nature of the studies upon which they were based.
Examples of scenes taken as repetition targets for comparison with data from former NASA manned spacecraft missions (years before) showed up the long term significance of orbital photography as documents for comparison of genetic features of environments, (e. g. landscape forming processes as lagoonal channel development of deltaic deposition).
There is an example of a Skylab photograph given in this publication.
One objective of the EREP programme was to determine which spectral combinations and individual wavelength bands are best suited to specific earth resources applications. This topic is not discussed in this publication.
Unmanned space remote sensing earth observation experiments such as NASA's ERTS–1 (Earth Resources Technology Satellite) which was launched 23 July 1972, nearly a year before Skylab.
The ERTS–1 satellite, since 1975 renamed "Landsat–1", made in two years 15 000 orbits. The total number of scene images received at U. S. data acquisition stations amounted to more than 13 000 and at stations in foreign countries to 49 000.
ERTS–1 showed the way to a new system and technique of orbital remote sensing. In comparison with manned spacecraft missions, such satellites are much less expensive, but very effective: continuous, long life (several years, for instance the ERTS surveying Skylab mission unexpectedly continued until 1975), world wide coverage due to its polar orbit, ERTS routinely provides high quality images of selected areas with its MSS (Multispectral Scanner) in several wavelength bands, from visible through infrared, from its 494-nautical-mile (915-kilometre) orbit.
With this orbital remote sensing system ERTS provided vertical images, giving a regular and constant scale, i. e. a constant sensor field of view and a repetition rate of 18 days for each scene. Geometric restitution is carried out by NASA's ground station computers with sufficient accuracy for small scale thematic mapping.
Although still in an experimental stage, as designed, this unmanned platform for orbital remote sensing will be developed into an operational system at some time in the future.
Some examples of interpretation and evaluation of picture information of ERTS images are given in this publication.

Technique of remote sensing systems

An understanding of the equipment used in orbital photography experiments and of the data characteristics is a prerequisite for the effective application of the data to practical problems.
The special techniques and characteristics of remote sensing systems such as colour photography, multispectral photography and MSS multispectral scanner imagery, are discussed briefly with the examples given in this publication.
Other remote sensing systems such as RBV-cameras (Return Beam Videcon), RADAR and thermography are in experimental or operational use in orbital remote sensing (e. g. in NOAA-satellites), but are not discussed here; digital data processing is making enormous strides in conjunction with specialized sensors such as electromechanical scanners, which can conveniently provide data in digital form. Specialized instruments are being developed for direct measurement of variables as different as sea state and ground moisture. In fact, parameters for which practical uses are as yet unknown can now be measured.
(For further information see handbooks and papers listed in the bibliography).

The interpretation of orbital remote sensing products, i. e. evaluation of the pictorial information

A number of very sophisticated statistical and computational methods are used for automatic recognition (e. g. computer-generated digital plots), to produce useful visual displays of the information recorded on the NASA ERTS-image data requisition and storage tapes. Certainly the development of electronic data processing techniques is important, the more increasingly numerous orbital remote sensing data become.

"Use of electronic data processing techniques is essential if one wants to classify and inventory this information within reasonable time." D. J. David, J. Deries and F. Verger, 1975, "Automatic cartography of ERTS remote-sensing data": Journ. of Brit. Interplan. Soc., vol. 28, pp. 624–628.

In addition, highly theoretical enhancement techniques, both photographic and digital, for thematic extraction, have been developed by scientists working on remote sensing methodology; these are not discussed herein.

"Several multispectral analysis systems are in various stages of research and development. Most of these are based on optical and photographic methods for combining the multispectral imagery for visual interpretation, using strong colour differentiation effects to display spectral reflectance changes. An almost endless variety of combinations of negatives, positives, internegatives and colours can be produced from the original photo records in arriving at an enhanced multispectral photo print. Thus, the data extraction process can become very complex, entailing special computer programs and elaborate and expensive printing and viewing equipment." D. S. Ross, 1969, "Colour enhancement for ocean cartography", p. 50 ff.: In "Oceans from space".

In this publication emphasis has been placed mainly on conventional data that can be used without extensive investment in specialized equipment and skills, but two examples of multispectral techniques and application are presented.

Colour or grey tone separation (as a type of photographic slicing technique) is discussed and used on the example of multispectral photography (coast of Texas), and to produce a small scale thematic map of tidal channels of a lagoon on the coast of El Salvador from ERTS-MSS-images. One example of false colour composite of ERTS-1-MSS imagery is presented in this book (compare cover and example El Salvador).

For the interpretation of satellite photographs and imagery, instruments as the Stereoscop from Oude Delft, the Interpretomat from ZEISS, Jena, and the Luftbildumzeichner from ZEISS, Oberkochen, have been used.

The Interpretomat is placed to the disposal of Prof. Dr. Gierloff-Emden, Department of Geography, Ludwig-Maximilians-Universität München, Laboratory of Remote Sensing (Fernerkundung, Luftbild- und Satellitenbildinterpretation) by the DFG (Deutsche Forschungsgemeinschaft) for research work.

Admittedly, a transition period is required. During this period, satellite data will be used in conventional ways – for example, with a strong emphasis on visual photointerpretation.

Individual visual interpretation of photographic and image products of orbital remote sensing is still the most valuable and effective method of evaluating the complex features and environmental systems of the earth's surface (though not for single parameters) because the human interpreter (if a scientist) of pictorial information is highly selective, in his approach.

This selection process is important, for not only do unnecessary data fill up storage bins but a superfluity of data masks out the important information, (process of filtering significant information from an abundance of data).

"Automatic data presenting by central computers of pictorial information is still in the far future".
S. A. Hempenius, 1976, *B. u. L.* p. 41).

This is certainly the case with complex coastal environments.

This publication presents the use of orbital remote sensing for coastal areas, for the margin of continents and oceans, and for the environments of lagoons, estuaries and bays: *The problems are always considered from the point of view of the earth and its environmental features, remote sensing is the tool, discussed in the case studies, presented herein.*

This study may be useful for further work on orbital remote sensing of coastal and oceanographic features, such as interpretation and evaluation of data available from NASA's Landsat–2 satellite, launched in January 1972, and expected from the proposed Landsat–C, to be launched in 1977.

Ground data acquisition: ground check and ground truth

"Two major study components are to be considered in any remote-sensing experiment: surface and remote data collection. Surface, or ocean truth, data are essential to remote-sensing research if specific properties of the environment are to be uniquely related to the characteristics of the remotely sensed image. Surface and subsurface measurements taken simultaneously with the remotely sensed data are particularly important where the environment is undergoing constant change as in coastal waters."
R. L. Le Mair, Photogrammetric Engineering 1973, pp. 928.

For optimal scientific use of orbital remote sensing and interpretation of pictorial information, as carried out with classical aerial photography, it would be necessary to check simultaneously the environmental parameters of the whole area in the sensor field of view, i. e. the scene.
This obviously cannot be achieved in producing a publication such as this using orbital photographs and images of the different satellite missions covering more than a decade: Gemini 5 launched in 1965 to Skylab 3 launched in 1975. Reports on ground check experiments are taken from the scientific publications, if available.
But most areas (all except three) used as examples in this publication are known to the author by personal visits or even as regions of scientific field work.
The author has studied coastal, and specifically inter-tidal, environments and aerial photography of these environments for 25 years, and published a basic study on the use of aerial photography concerning the coastal environment in 1961:

H. G. Gierloff-Emden, 1961, Luftbild und Küstengeographie am Beispiel der deutschen Nordseeküste: Landeskundliche Luftbildauswertung im mitteleuropäischen Raum, *Schriftenfolge d. Inst. f. Landeskde. i. d. Bundesanst. f. Landeskde.*, vol. 4, Selbstverlag d. Bundesanst. f. Landeskde. u. Raumforsch., Bad Godesberg, 118 p.

Geometric accuracy of orbital photography and imagery

Since in the examples given in this publication, the problems are always considered from the point of view of the earth and its environmental features, and the orbital photographs and images provide the basic data for evaluation, in the sense of visualising coastal and marine environmental phenomena and processes, interpretation is a qualitative process (i. e. interpretation maps are mostly produced in the same geometric pattern as the orbital photographs). For this purpose geometric restitution and rectification are unnecessary.
In the case of thematic mapping of coastlines and inter-tidal areas of a lagoon from ERTS images the geometry accords with the accuracy as given by NASA's ERTS image material.
(Compare this chapter. For details: NASA – ERTS – Data Users Handbook, and G. Konecny, 1975, Approach and status of geometric restitution for remote sensing imagery: *B. u. L.* 1, 1975, p. 2 ff.)

Spectral aspects

Potential use of multispectral scanner bands for recognition of target features (marine and ocean surveys)

Experiments were carried out with NASA's Skylab missions (Skylab Earth Resources Experiment Package, EREP), using 13 band multispectral scanners (band 1, wavelength from 0.41 micrometre to band 13, wavelength to 18.5 micrometre).
"A preliminary analysis of the capabilities of the individual multispectral scanner channels, based in part on published investigations that used discrete wavelengths similar to those of the scanner, indicates that some channels are more useful than others in detecting and defining key natural and cultural phenomena", (as reported in NASA Skylab Catalogue, table 5).

for:	waste disposal in coastal waters penetration, turbidity, sea ice acquisition of maximum underwater detail varies with turbidity	band 3, wavelength 0.49–0.56 micrometre band 4, wavelength 0.53–0.61 micrometre
for:	shoreline configuration, wetland mapping land water contrast	band 6, wavelength 0.64–0.76 micrometre band 6
for:	land water discrimination	band 7, wavelength 0.75–0.9 micrometre band 8, wavelength 0.9 –1.08 micrometre

Coastal boundary surveys

Water is imaged as almost pure black on the infrared photograph and as black and white infrared of multispectral photography and Multispectral Scanner MSS band 7, whereas the beach occurs in light grey or white. Due to this effect the separation of beach and sea (or lagoonal water) is depicted with high contrast and can be used for coastal mapping, especially for coastal boundary surveys and rights of ownership. This is important, since the coastal zones of the earth are being given new emphasis due to the importance of exploitation in particular of mineral resources and fisheries.

Detection and interpretation of colour on orbital photography and imagery

Colour is the bearer of information on satellite colour photographs and images.

"Except for shorelines and similar features which vary sharply in colour or form the surrounding terrain, the oceanographer – hydrographer is primarily concerned with recognizing and interpreting the meaning of very small changes in colour hue and colour saturation, whether it be in the colour photograph or the multispectral record."

D. S. Ross. 1969, Colour enhancement for ocean cartography, pp. 50–63: in Oceans from space.

When interpretating, objects are described by their appearance in colour and tonal variations. Colours are physically identified in terms of wavelength and intensity, and photographic films are sensitive in the sense of the wavelengths. But for perception by the human eye, the colour of surfaces is much more complex. The photographic film emulsion may contain, and the human eye may distinguish, subtle variations of colours. Homogeneous areas (small patches only on the photographs to be interpreted) of bright saturation and hue lead to what is termed colour on the photographs.

Saturation is the attribute that determines the degree to which a colour differs from white of the same brightness. *Hue* is the attribute that makes it possible to classify colours as red, yellow, green, blue or intermediaries of these. Differences in hue correspond to differences in wavelength. Visual differentation of colours is somewhat subjective, but can get a degree of objective comparability by the use of a colour code. In 1929 the Munsell Colour Company introduced the "Munsellscale", i. e. Munsell Colour Solid which assist basically in specifying the appearance of a colour by comparing it visually with a set of standard colour chips or samples. This system defines 100 hues of equal perceptable differences, associating the colour name with the visual sensation. More advanced studies may use instruments for colorimetry or colour sensitometry. Basically the exposed film is the bearer of colours depending on either the film in the camera or the film on which the colour composite is exposed from 3 or 4 different black and white films, each a product of a certain spectral band of multiband scanner imagery as MSS.

The three factors of *hue*, *saturation* and *brightness* exhaust the degree of a film colour. If two film colours are separately equated for the factors hue, saturation and brightness the colour will appear identical. The problem is that each developed copy of colour film is a little bit different and the more *copy generations* are produced, the more the details get lost. NASA usually delivers copies of third or sometimes second copy generation. These may be still fairly good, but with each photographic process followed from then on the colour of the picture will vary. This is the reason why in different publications the same photograph may appear so different in colour (mainly in factor hue).

Printed colour pictures

It is noteworthy that billions of dollars are paid to obtain the basic material of orbital photography and imagery, but that it is extremly difficult to get money for very good colour prints and the best available, such as those needed for publications of Gemini photographs, are only realised in some very few NASA publications. The current process of printing uses only four standard industrial colours, e. g. red, blue, yellow and black, whether using a screen process or autotype; a technique of overprinting the four colours each with a very fine grain gives the human eye the impression of a wide range of colours. The character of these prints can easily be detected by using a magnifying lens, (see also the examples in this publication). These printed pictures don't have the same quality as the original genuine film pictures which were used for interpretation and moreover differ in hue, saturation and brightness, i. e. the demonstration material is different from the material used for scientific interpretation.

For detailed information see:

J. T. Smith, jr. and A. Anson (ed.), 1968, Manual of colour aerial photography, 1. edition: American Society of Photogrammetry, Falls Church, Virginia.

American Society of Photogrammetry, 1969, New horizons in colour aerial photography. A seminar sponsored by the American Society of Photogrammetry and the Society of Photographic Scientists and Engineers: 423 p. Falls Church, Virginia.

S. Wenderoth and E. Yost et al., 1974, Multispectral photography for earth resources: Remote Sensing Information Center, Greenval, N.Y. First publ. for the NASA, Contr. no. NAS 9-11188 and subject to Clause 90, Rights in data.

Standard:

The Munsell book of colour, 1929: Baltimore.

The colour of the sea

"Identifying different oceanic environments by the colour of the sea was a common practice among seafarers of an earlier day who left us the legacy of such names as the Azure Sea, the Black Sea, the Red Sea, the Yellow Sea, the Kuroshio or Black Current, the White Sea, and the Vermillion Sea. In the Easter season, the Bay of Panama frequently turns red due to the spring blooming of a microscopic floating plant – and the early Spanish explorers referred to this as lachryma christi. Christopher Columbus, as was the custom of his day, used sea colour as a means of finding his way about over the trackless ocean."

G. C. Ewing, 1969, "The colour of the sea", p. 46 in "Oceans from space".

With the advent of satellites, observation of sea colour may again find a use but with a new objective. The spectacular colour photographs taken by the Gemini, Apollo and Skylab astronauts are full of information of significance to students of oceanography from space.

Atmospheric and oceanic phenomena can have similar appearences on orbital imagery, due to dependence on colour, therefore sea colour must be taken into account. Ambiguity of the information as detected but not separately identifiable on the image can exist.

Colour difference can be caused also by atmospheric phenomena. Light and colour vary with daylight (time of the day), sun elevation, sun azimuth in relation to the camera exposure axis direction, and cloud types and distribution, as well as on oceanic features such as sea state and turbidity of the water masses. Hydrographic characteristics are good indices for different water masses.

It has been found that at heights above 3 000 m, atmospheric effects do not increase and thus scene radiance values in blue spectral records are not degraded further. This is important for orbital photography.

Orbital remote sensing gathered information from back scattered and reflected daylight and sunlight, both out of the sea and from the sea-surface, on the exposed film or sensor, produces colour as a spectral reflectance.

This colour of the sea appears different to the spaceborne system from the colour with which it appears to the eye or camera for instance through a glass window of a boat.

The colour of the sea

The colour perceived by a human eye observing the sea, if avoiding the reflections of the clouds and the sky, is determined by the spectral energy distribution of that part of the light which penetrates the surface of the sea and then, according to how it is dispersed and reflected from particles suspended in the water, returns to the surface as luminance from the water body. The colour changes from deep blue in tropical and sub-tropical seas to bluish-green in higher latitudes; to green in regions of upwelling waters and shelf areas, and to the discoloured green of very turbid coastal waters.

The optical properties of seawater determine the depth of penetration and the spectral composition of the light in the sea. They therefore influence the temperature of the water, the colour, the assimilation limit of phyto-plankton, (primary production) and thus the basis of all marine life. These properties are therefore crucial and may be listed as follows: reflection, refraction, dispersion and absorption – all of which determine the dispersal of light rays in the sea and affect the attenuation.

The top ten metres of seawater in the open ocean absorb between 33 and 80 percent of the blue light, depending on the clarity of the water.

The sea surface has a very limited albedo for the incident light rays. At an altitude of the sun of more than 25° it is less than 10 %. The refraction of light in water depends on temperature and salt content. For this reason it can be simply used to determine the salt content if the temperature is known. Dispersion and absorption determine the total absorption, the so-called attenuation of the light rays. Both depend to a large extent on the wavelength of the light: dispersion increases with decreasing wavelength of light and absorption with increasing wavelength of light, so that absolute minimum attenuation occurs at 0.47 μm, i. e. in the blue of the chromatic spectrum. Theoretical deduction and practical observation show that for pure seawater in sunlight the maximum amount of light returning from the water to its surface is gained at a wavelength of 470 μm. If the sky is clouded, this maximum is shifted some μm toward the longer waves in the spectrum. Changes of this fundamental colour of the sea are caused by various influences. (From G. Dietrich, and theories on the colour of seawater from Bunsen 1847, Raman 1922, and Kalle 1938).

The colour of seawater corresponds to that wave range of light at which the ratio of back dispersion and absorption of the light entering attains its maximum. This is the situation at the specified wavelength of 0.47 μm (from G. Dietrich, Allgemeine Meereskunde, and Geophysik-Lexikon). For this reason the colour of pure seawater is an intense cobalt blue as is typical for the subtropical regions of the oceans and as is shown predominantly in oceanic waters in certain pictures. Colourless suspended particles intensify the dispersion irrespective of the wavelength and thus lighten the blue seawater colour. Therefore the reproduction of submarine relief features by aerial or satellite photographs and images depends on the nature of the surface water masses. Moreover, the angle at which the sun strikes the water is also very important for the evaluation of aerial of satellite photographs and images.

The influence of the atmosphere on remote sensing measurements ultraviolet; visible and infrared region: ed. by J. B. Farrow, European Space Agency, Contr. 1837/72.

J. E. Arnold and A. H. Thompson, "Apparent coastal water discolourations from space, oceanic and atmospheric." pp. 82–91, in "Oceans from space".

N. G. Jerlov: Optical Oceanography, Elsevier Publishing Company, 1968, Amsterdam.

Oceanography from Space: ed. by Gifford C. Ewing, The Woods Hole Oceanographic Institution, Woods Hole, Mass. 1965.

Peter C. Badgley, Leatha Miloy, Leo Childs: Oceans from Space. Gulf Publishing Com., Houston, Texas. 1969.

Transparency

For interpretation and detection, coastal phenomena of high importance are the transparency of water masses and the optical behaviour of turbid water masses.

Intensity of light decreases with depth of the sea. Coastal waters may appear clear to the eye, but they are always turbid in comparison with oceanic waters.

Transparency of the sea is usually described and measured by the depth of water at which a white disc., (the Secchi disc.) is visible from above. This is a measure of light penetration into the ocean.

This definition of transparency differs from total penetration as measured by a diver or from a submarine at depth; total penetration is greater than transparency as measured by Secchi disc by 10 to 100 times.

Illumination at increasing depths follows an approximately exponential law and depends on many parameters.

Turbidity

Oceanic and coastal waters are a mixture of solution, colloidal phase, suspension and turbidity of inorganic and organic particles and yellow colour substance, i. e. turbidity has a very complex optical physical character.

Definition: presented by the U. S. Geological Survey; (1964)

"*Turbidity* is the optical property of a suspension with reference to the extent to which the penetration of light is inhibited by the presence of insoluble material. Turbidity is a function of both the concentration and particle size of the suspended material. Although it is reported in terms of parts per million of silica, it is only partly synonymous with the weight of sediment per unit volume of water."

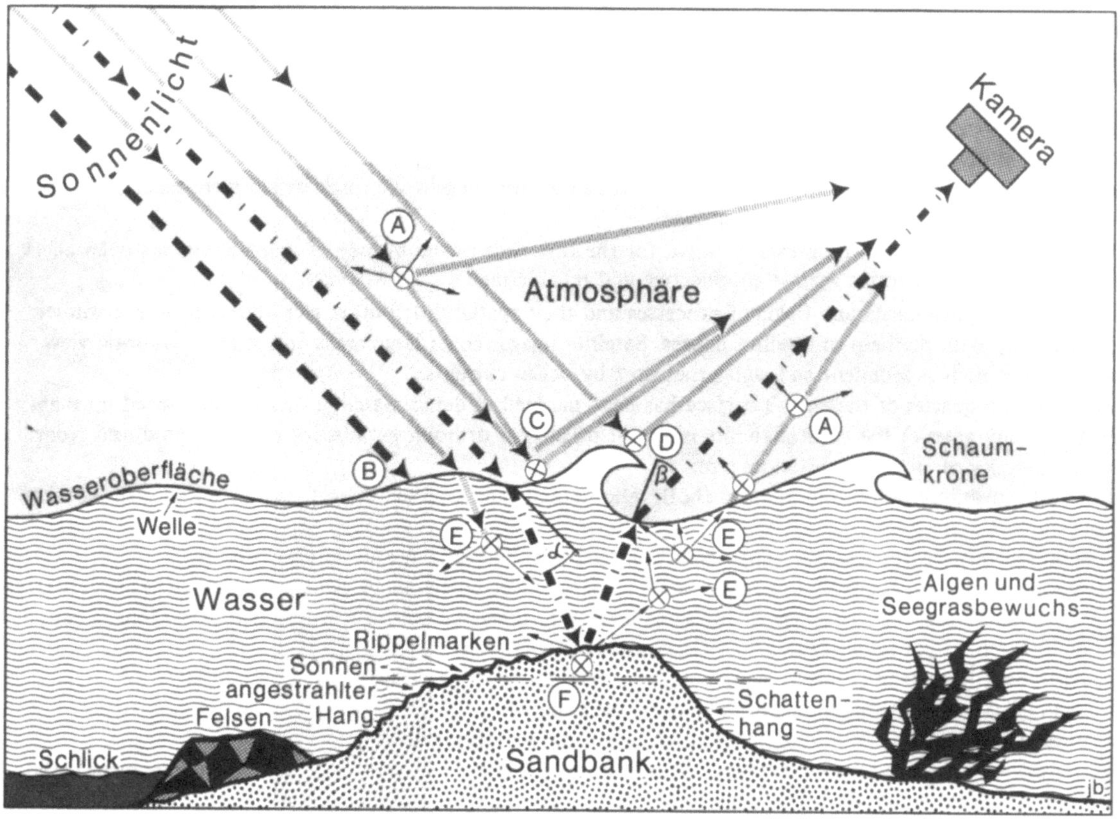

Scattering, absorption, and contrast attentuation of camera image flux by atmosphere and water

(as occurs when the camera exposure axis is in juxtaposition (opposite) to the sun's azimuth)
Drafted after a design by D. S. Ross, 1969, p. 52: in P. C. Badgley, L. Miloy and L. Childs, Oceans from Space.

A – Atmospheric haze: blue scatter
B – Absorption of red light
C – Surface reflection of sun & haze
D – Whitecaps
E – Reflection, absorption & scattering in water
F – Diffusion of light from bottom

Translation of terms:

Sandbank	= sand bar	sonnenangestrahlter Hang	= sunlit slope
Felsen	= rock	Schattenhang	= shadow slope
Schlick	= mud, ooze	Welle	= wave
Rippelmarken	= ripples	Schaumkrone	= white cap

The different slopes of a sand bar, as a feature of submarine topography, may reflect different light intensities, depending on their exposure to sun's rays: sunlit slope and shadow slope, although both slopes are located at the same depth, i. e. covered with the same water column.
The pictorial information on the photograph which can be detected (grey-tone or hue, colour tone) depends also on the sun's azimuth to the slope and the sun's elevation angle to the water surface, the surface of the bottom topography and to the direction of wave propagation and sea state.

Application of satellite photographs and images for the production of thematic maps

Space technology provides satellite images of the earth's surface. For the first time an objective, direct optical generalization of the visible topography, can be carried out with the help of these images. Up to now small scale maps were drawn by subjective generalization from large scale maps which were themselves abstract representations of the earth's surface.

"Let me conclude by repeating that ... now, for the first time in the history of mapmaking, man can check the generalization of forms against an objective and reliable model." C. Koeman, 1970.

Earth surface phenomena and fields of processes and their spatial distribution can be recorded and studied scientifically with the help of satellite images. Satellite images cover large areas and allow a synoptic view of processes such as sediment and water transport by ocean currents.

Only about a quarter of the earth's surface has been mapped in detail. Satellite imagery can speed up mapping of many areas of the earth. Satellite image maps such as orthoimage mosaics could be produced from colour mosaics of areas of substantial size.

The basic advantage of satellite pictures (both, photographs and scanned images) compared with those taken from aircraft are:

- simultaneous recording of very large areas (up to 200 x 200 km) in only one picture;
- synoptic pictures of global phenomena, such as ocean tides on flats and lagoon coasts, and sea-ice coverage
- "automatic" repetitive coverage of the same area from the same platform within shorter or longer periods of time (this from satellites as ERTS (Landsat) or NOAA);
- Earth-oriented satellite imagery can be used for the following projects:
- regional research, mapping of morphological landscape forms;
- present day geological-hydrological research, e. g. tidal sediment transport;
- analysis of processes by studying and dating relevant parameters in a single image, and comparison with imagery taken on different dates (surveillance).

Thematic maps of the scales from 1 : 250 000 to 1 : 500 000 are called "small scale" or "medium scale" in different scale classification systems.

Interpretation

The interpretation is *qualitative* as it concerns the recognition (detection and identification) and localization of objects and their classification into categories.

The analysis is *quantitative* as it concerns comparison with former interpretation of pictures or maps from field work, so that measurements of the development of phenomena can be made.

The *evaluation* is subjective because it depends on the competence of the interpreter and is selective depending on his choice of the phenomena to be evaluated.

The *pre-information* of the scientist provides a basic knowledge of the specific subject of study, including its geography, oceanography, cartography, morphological mapping, knowledge of the image scene in regard to the geomorphological type (here: the coastal region) and knowledge of the specific geographical region together with details of its environmental features.

Example: The existence of a light grey strip on the picture is detectable against dark strips (boundaries by linear elements). The strip is identified as a beach ridge on a barrier beach.

Selection of relevant information on the scientific problem is obtained from the whole range of information obtained from a satellite image. The amount of detailed information provided by the satellite is much larger than that specifically needed for the purpose. This step in the interpretation corresponds to a filter, whose transmission factor depends on the pre-information of the scientist (i. e. selection based on previous knowledge).

For genetic description, information can be gained only indirectly from the picture by analyzing phenomena that provide clues to certain processes, e. g. sediment clouds in water sediment transport.

Application of ERTS-1 Multispectral (MSS) Imagery to small scale map cartography of the coastal zone of El Salvador

The four-channel imagery of the multispectral scanner in the ERTS–1 satellite has been applied to the production of a small scale thematic map (1 : 300 000) of a lagoon area of 50 km length by 15 km width. Details from the various levels of the terrestrial, inter-tidal and marine environment, such as vegetation in the inter-tidal zone, could be extracted by analysis and integrated interpretation of ERTS-MSS- images with sufficient accuracy for small scale mapping. Elements of the landscape (environmental features) which are below the minimum resolution of the imagery (but known from fieldwork) were not mapped. It was possible to define the boundaries of the various vegetation and tidal zones. Mangrove could be differentiated from the terrestrial vegetation and also from water areas on the same inter-tidal level along the tidal channels and inlets.

The interpreted images were produced by enlarging each of the four ERTS-MSS (Multispectral Scanner) image NASA copies three times. The original scale of the NASA copies of ERTS-MSS images is 1 : 1 000 000 i. e. 7.3 inch image size on films of 9 inches square. On these, the images themselves measure 185 mm, i. e. 1 mm on the image is equal to 1 km on the ground.

The image interpretation of the phenomena and the objects of environmental features of the coastal region was carried out using the four ERTS-MSS spectral bands.

The interpretation of geomorphological-oceanographic features is based on information obtained during an intensive ground check carried out by the author in 1954 and 1955. At that time the morphological reference map was also prepared, based on fieldwork and aerial photographs on a scale of 1 : 30 000. This ground-checked map, prepared from large scale aerial photographs, was compared with the new map from ERTS-MSS images. It has been concluded that the ERTS multispectral scanner imagery is quite suitable for the given task of preparing small scale maps of the coastal region on which the boundaries of each landscape "level" (= "Stockwerk") and vegetation class are defined and determined. On this map, the area of these features can be measured. (On a map of the scale of 1 : 500 000, an object of 50 m wide is equal a line of 0.1 mm wide).

For the production of such a thematic map, all four ERTS-MSS bands were found to be useful. Bands 4 and 5 provided water transparency in shallow areas, and bands 6 and 7 contrasted the terrestrial level with the water surfaces.

The ERTS-MSS images have been used to map the surface distribution of the vegetation types in the shore environment. Outlining areas within a continuous surface on the map involve a decision process by the researcher. In the present study, the limits of the various habitats within the coastal zone were established on the basis of local knowledge of the environment including its features and processes, e. g. lagoons and the tidal zone.

The boundary lines and boundary zones have been investigated on the basis of their presentation on the four ERTS-MSS bands. Small features which are below the minimum resolution of the satellite imagery, or which due to cartographic limitations, cannot be shown on the map, and have been rejected during drafting.

The MSS prints consist of 9 inch square images. On these, the images themselves have a side-length of 185 mm, i. e. scale 1 : 1 million. These pictures were produced by NASA by enlarging the 70 mm films. The following interpretation is made from images on 9 inch films. Some specific areas of these images were enlarged three times, i. e. to a scale of 1 : 300 000 for this work carried out by the author.

Photographic automatization in the compilation of thematic maps.

The MSS satellite image delivers more information than a panchromatic photograph. In this case the mapping of the inter-tidal landscape levels in the coastal region for a thematic map is made possible due to the various MSS-bands.

Analogous conclusions are used. Because of this and due to the scientist being able to apply previously acquired information for this interpretation, the use of "automatic" analysis is only applieable for specific problems and is only an advantage over traditional compilation methods for certain quite subjective interpretation.

In the present example the mapping of the water area of the lagoon could be done by photographic selection, i. e. automatic analysis with the aid of ERTS-MSS band 7.

Certain areas on land must however be omitted as they appear black but are known not to be water (e. g. young, rough, black lava field), because this is ambiguous information.

The category of objects shown by grey tones and outlines in the photo images is not necessarily identical with categories of certain landscape features. Differences in grey tones are caused by the differem reflection characteristics of various environmental features such as water, sand, rock, forest or cultivated land, but depend also on the sun's elevation and azimuth.

After analysing the geomorphological units of the environment, the objects identified must be classified to constitute a morphological legend, e. g. in coastal area, bays and lagoons of El Salvador, classification was based on previous field work.

The Resolution

The inherent information contained in an image, depends, apart from contrast, upon resolution. One can distinguish between ground resolution, the object size given in metres (minimum visible), and image resolution. Detail detect-ability, meaning the size of the smallest object that can be recognized on the photograph by grey tone or texture, depends on both categories of information. (Photographs contain both, grey tone and texture; scanner-images do not show texture, due to the line scanning technique used).
Resolution of the eye is given as 5 lines per mm (without magnifying lens). The image should have at least 10 lines per mm resolution in order that maximum advantage can be taken of the scope of the information.

Ground resolution depends on the shape of the object; and on the contrast ratio against the surrounding area: point, line area and surroundings; e.g. small sand island, beach line, barrier island. When interpreting aerial and satellite images, it should be noticed that linear objects can be detected even when their width is less than the resolution of small patches. In this case the relation patches: line is 5 : 1. This relates to tectonic-lithologic lineaments, joint and morphological lineaments, valley segments, as well as boundaries of vegetation – land – water and cultural-geographic objects (canals, streets).
The signals of small objects which contrast poorly to their surroundings are overshadowed and covered by the signals of the surroundings. Target elements oriented parallel to the scan direction of the ERTS-MSS-images might be obscured due to the scanning process. The example given here (an ERTS-MSS image interpretation of the coastline of El Salvador) concerns beach ridges on a barrier island.
Gemini and Apollo photography has ground resolution from 50 to 100 m, Skylab photography from 20 m to 40 m.
The ground resolution of the objects should be in a significant relationship to the scale of the map to be plotted. To produce a map on a scale of 1 : 1 million, a ground resolution of 50–100 m for detailed mapping is necessary, 100–500 m is however sufficient for an exploratory informative map. Wobber (1972).

Image resolution refers to camera photographs and is given in lines per mm on film. The resolution is a function of contrast reproduction (Heynacher and Köber, 1964).
The image resolution in a scanner system is pre-determined by the number of the scanner lines. Image resolution for ERTS-1-MSS images is given as 300 m for objects of unit plain surface (so called symmetrical objects) and is better for line shaped objects. Cameras produce 50–70 lines per mm, black-white films 50 lines/mm, colour films 70 lines/mm.

Image resolution, enlarging of images and mapping

The scale of the ERTS images on 70 mm film is 1 : 3.3 million and they have a theoretical image resolution (independent of the scanner-system) of approximately 10 lines/mm. The result of photographic, i. e. optical enlargement is that the image has a smaller resolution than the eye (10 lines/mm).
According to the condition $Ma = r_p/10$ (wherein Ma = possible enlargement of photo to map scale, r_p = original image resolution in lines per mm), the present enlargements deliver less information regarding their resolution than it would be possible to show in a map. This is justifiable in the present case because the aim of the map is mainly the representation of limits of only 3 object categories of environmental features of the coastal complex:
The three object categories of the coastal complex
I. barrier peninsulas or islands with beaches above high water level (terrestrial)
II. inter-tidal flats with mangrove vegetation (inter-tidal or drying zone)
III. marine areas of the bay and lagoon below low water level (lagoonal or marine).

The branching of the lagoon's tidal channels can be detected as linear elements on the image in scanner band 7 with best contrast, and can even be recognized beyond the areal resolution limit, so that in this case the mapping of lagoons with tidal channels from satellite images is efficient. The contours of an object mapped in the example of the coastline of El Salvador and map-scale naturally appear simplified because below a certain value no contrasts can be detected. In this case the branching of the vegetation limit of the mangrove appears simplified (smoothed) and tidal channels whose branches are narrower than 60 m (exceptionally 30 m) and a lot of existing small ditches are not visible and therefore remain unmapped.

The coastal environment

Classification of coastal environments including tidal zone terminology

The topographic and hydrographic features of the coastal environment are classified by water levels which determine three distinct zones.

The extraction of informations from satellite photographs and imagery with the task of classifying it into these three categories has been one of the main work and design done by the author in this publication.

In this publication, the legends used in the interpretation maps are as follows:
I. above High Water level: as terrestrial
II. between High and Low Water level: as the inter-tidal zone
III. below Low Water level: as lagoonal and marine.

This is a very general classification using the different water levels in tidal or even in non-tidal areas. Differentiation due to distinct water levels as defined under "Tides and Currents" in:
"Symbols and Abbreviations used on *Admiralty Charts*", Hydrographic Department, United Kingdom, Chart 5011, ed. 1973, is not only unnecessary for small scale interpretation maps, but is often impossible due to lack of tidal data: most of the lagoons and estuaries shown in these orbital photographs and images do not have a network of tide gauges of sufficient density to allow differentiation of the 12 tide-caused water level definitions which could be used in the interpretation maps of this publication. In the work: "Problème de photogéographie et de représentation cartographique du littoral Atlantique Français", presented by R. d'Hollander, as an example of exact and detailed classification and differentiation for the mapping of charts of coastal and inter-tidal areas, a map in large scale in 1 : 20 000 of the region of the environment of the Ile d'Oléron is given.
The terms "tide" and "tidal", as described for use and definition of sea level fluctuations produced by the gravitational attractions of the moon- and sun-astronomic tides, are used in this publication (as in many older publications on coastal environments) also to describe levels of the environment with mostly small but important vertical extension but causing areas of extensive horizontal extension, as for instance tidal flats or tidal marshes; these are called tidal zones or inter-tidal zones.
"There is rather general agreement on the boundaries of the tidal zones, but there are differences among some authors on the upper limits of "intertidal" and "tidal flat"."
Reineck, 1975, North Sea tidal flats: in Ginsburg, R. M. (ed.), 1974, Tidal deposits, a casebook of recent examples and fossil counterparts: New York, Springer.
Tidal zone terminologies are differentiated sometimes as:

"above High Water	as supralittoral or supratidal zone
inter-tides	as littoral
below Low Water	as infra- or sublittoral or subtidal zone"

M. Gary, R. McAfee and C. L. Wolf (eds.), 1972, Glossary of geology: Am. Inst. Geol. Washington, 857 p.

The classification used in this publication defines the differentiation of the coastal and marine environment; this use is feasible for interpretation of orbital remote sensing, both photography and imagery.
Depending on the ground resolution, small scale features (up to 50 m) are not discernible on these photographs and images (except linear subjects on Skylab Earth Terrain Camera photographs, where tidal channels of 10 m to 20 m width can be detected and identified), so there is no sense in using these small scale features of coastal environments for classification.

Dolan, R., B. Hayden and M. Vincent, 1975, Classification of coastal landforms of the Americas: *Zeitschr. Geomorphol. N. F.*, suppl. vol. 22, pp. 72–88.
McGill, J. T., 1958, Map of coastal landforms of the world: *Geogr. Rev.* vol. 48, p. 402–405.
McGill, J. T., 1959, Coastal classification maps, a review: 2nd Coastal Geogr. Conf. 1–21.
McGill, J. T., 1960, Map: in W. C. Putnam et al., Natural coastal environments of the world, Los Angeles.
Jonin, A. S., P. A. Kaplin and V. S. Medvendev, 1961, Classification of the world sea shores: Proc. Oceanogr. Comm. Academy of Sciences USSR, 12.
Nordstrom, C. E., 1971, On the tectonic and morphologic classification on coasts: *Journ. Geol.*
Shepard, F. P., 1937, Revised Classification of Marine Shorelines, *Jour. Geol.*, vol. 45, no. 6, pp. 602–624.
Shepard, F. P., 1952, Revised Nomenclature for Depositional Coastal Features, *Bull. Amer. Assoc. Petroleum Geol.*, vol. 36, no. 10, pp. 1902–1912.

Environment, and the sciences: Geography, Oceanography and Remote Sensing

Though oceanography used to be a branch of geography as a global "earth science", it is of some interest to think on the former use of this term "environment". The term "environment", now used so widely has a long history in science and has very different uses in different countries. Even a shortened description of the use and definition of the term cannot been given here but some quotes follow. For a detailed study, see G. Fuchs (1966).

"Environment" was, before 1900, a term of philosophy, then used by geographers instead of "physical feature". "Environment" was used in the sense of factors and conditions of natural landscapes on the human being in the sense of determinism, called "environmentalism". The relationship of man to the environment was then called "human response" (later including the social behaviour of man); this all pertained to theoretical work.

It was Carl O. Sauer of Berkeley, California, who, since the twenties of our century has set up field studies for detailed observations on the "natural environment", as an interplay of influences between man and landscapes.

"It is no detraction of the field of geography to insist that a very great amount of precise systematic and analytic work needs to be done before time is ripe for the more brilliant synthesis of broad conclusions. As yet we have hardly begun to state to ourselves the problems of the field". P. 19.

Sauer, C. O., 1924: The Survey Method in Geography and Its Objectives. In: Annals. Assoc. Amer. Geogr., 14, 1924, pp. 17–33.

The term ecology was at the same time adapted and used by geographers to explain the "environment", including the impact of man on the natural landscape.

The scale of man's impact varies from the molecular to the global (see Dansereau, 1966, pp. 449–452).

"The works of man express themselves in the cultural landscape. There may be a succession of landscape with a succession of cultures. They are derived in each case from the natural landscape, man expressing his place in nature as a distinct agent of modification."

Sauer, C. O.: The Morphology of Landscape. In: Univ. Publ. Geogr., 2, 1925, S. 19–53.

Th. R. Detwyler wrote in his book "Man's impact on environment" (1971) p. XI

"Man's Impact on Environment substantiates the wide spectrum of environmental changes wrought by man, focusing on major processes of change, immediate and extended effects on the environment, and trends in time and space of processes and effects."

Detwyler points out the widening gap between change and understanding: the change of ecosystems of landscapes is going on fast, the details of many ecosystems and the change due to man's impact on the environment are not well understood, although there is in our time an accelerated change. The question is, can we overcome our ignorance and apathy concerning our environmental actions soon enough?

"The environment is artificially divided into separate components for study.These environmental aspects generally coincide with academic disciplines." "There remains a grave paucity of interdisciplinary, broadly integrative studies of the environment, in part because of increasing specialization in science. The greatest, but still only partial, exceptions to this situation are found in the fields of ecology and physical geography." "No single scientific movement has yet integrated these features, and hence there is no explanatory or predictive environmental science." R. Detwyler, 1971, pp. 2–5.

The term "environment" is introduced as a term in the sense of features, patterns and processes of a certain geographic, geological or oceanographic characteristic of nature, often including the impact of man on ecological systems.

The scientific work of the sixties, and even more so of the seventies of our century turned out numerous publications in which the term *environment* is used in this sense (see bibliography of some books filed up here).

In this publication the terms "environment" and "environmental features" are used to describe and to clarify certain phenomena, parameters, processes and patterns of the coast including the terrestrial (continental), inter-tidal and marine (oceanic) zones of the interface of land and sea, in certain geographic regions. The term "littoral" is not used here because it has a specific but different meaning in coastal geomorphology, sedimentology and biology.

With the development of remote sensing techniques, both aerial and satellite photography and imagery, a new tool is available to handle the problem of the environment in our time.

In this publication, a number of selected examples of orbital remote sensing of the coastal environment, specifically of lagoons, estuaries and bays, will be presented. The importance of remote sensing to the present attempt to stimulate and update our knowledge of the "environment", has been stressed.

Remote sensing of the environment is also discussed and presented as an interdisciplinary topic on geography, oceanography and remote sensing. Orbital remote sensing can be an useful tool when responding to the question: How do processes occurring at different scales in space and time interact?

Bibliography concerning the "Environment" as a term in the sense of "man and environment" and "ecosystems" of the earth

Dansereau, P. (ed.), 1970, Challenge for survival, land, air and water for man in megalopolis: Columbia Univers. Press, New York and London, 235 p.

Demoll, A., 1954, Bändigt den Menschen, gegen die Natur und mit ihr. Der Mensch ändert seinen Lebensraum: Bruckmann Verlag, München, 306 S.

Detwyler, Th. R., 1971, Mans impact on environment: McGraw Hill, New York, 731 p.

Fels, E., 1965, Der wirtschaftende Mensch als Gestalter der Erde: Frunckh, Stuttgart.

Fuchs, G., 1966, Der Wandel zum antropogeographischen Denken in der amerikanischen Geographie. Strukturlinien der geographischen Wissenschaftstheorie, dargestellt an den vorliegenden wissenschaftlichen Veröffentlichungen 1900–1930: *Marburger Geogr. Schrft.*, Diss. (D 4), Geogr. Inst. Univers. Marburg, no. 32.

Glacken, C. J., 1956, Changing ideas of the habitable world: in W. L. Thomas jr. (ed.), Man's role in changing the face of the earth: The Univers. Chicago Press, pp. 70–92.

Marsh, G. P., 1864, Man and nature, or, physical geography as modified by human action: Scribners, New York and Sampson Low, London, 560 p.

Nicholson, M., 1970, The environmental revolution, a guide for the new masters of the world: McGraw Hill, New York, 366 p.

Ray, G. C., 1975, A preliminary classification of coastal and marine environments: International Union for Conservation of Nature and Natural Resources, Morges, 26 p.

Russell, R. J., 1956, Environmental changes through forces independent of man: in W. L. Thomas jr. (ed.), Man's role in changing the face of the earth: The Univers. Chicago Press, pp. 453–470.

Sauer, C. O., 1925, The morphology of landscape: *Univers. Publ. Geogr.*, no. 2, pp. 19–53.

Sherlock, R. L., 1931, Man's influence on the earth: Home University Library of Modern Knowledge, London, 256 p.

Thomas, W. L. jr. (ed.), 1956, Man's role in changing the face of the earth: Univers. Chicago Press, Chicago, 1193 p.

Wilson, B. R. (ed.), 1968, Environmental problems – pesticides, thermal pollution, and environmental synergisms: J. B. Lippincott, Philadelphia, 183 p.

Bibliography concerning the "Environment" as a term in the sense of features, pattern and processes of a certain geographic, geological or oceanographic characteristic of nature
(often including the impact of man on ecological systems)

Dolan, R., B. Hayden and G. Hornberger et al., 1972, Classification of the coastal environments of the world. I. The Americas: Dept. Environm. Sci. Univ. Virginia, Techn. Rep. 1, 86 p.

Hedgpeth, J. W., 1957, Classification of environments: *Geol. Soc. America*, Mem. 67, vol. 1, p. 17–27.

Nelson, B. W. (ed.), 1972, Environmental framework of coastal plain estuaries, based on presentation at the symposium on estuaries, 18th annual meeting of the southeastern section of the Geological Society of America, Columbia, South Carolina, April 10 and 11, 1969: *Geol. Soc. America*, Mem. 133.

Putnam, W. C. et al., 1960, Natural coastal environments of the world: Univers. of Calif., Los Angeles.

Ranwell, D. S., 1972, Ecology of salt marshes and sand dunes: Chapman & Hall Ltd, London, 212 p., 83 fig.

Reineck, H. E. and I. B. Singh, 1973, Depositional sedimentary environments: Springer, Berlin-Heidelberg, 439 p., 579 fig.

Schäfer, W., 1972, Ecology and palaeoecology of marine environments: Oliver & Boyd, Edinburgh, 568 p. (A remarkable synthesis of functional morphology, life habit, death and preservation of marine animals in North).

Nature and Resources, Department of Environmental Sciences and Natural Resources Research, Unesco, 7 Place de Fontenoy, 75700 Paris (France). (*Nature and Resources* is the official bulletin of: the Man and the Biosphere Programme, the International Hydrological Programme, and the International Geological Correlation Programme).

Phenology and Seasonality Modeling. Edited by Helmuth Lieth, Berlin, Springer-Verlag, 1974, 464 p. (Ecological Studies, vol. 8)
Most of the chapters of this book were first presented as papers at a symposium held in Minneapolis, in August 1972, during the twenty-fifth annual meeting of the American Institute of Biological Science. Seasonality problems are discussed within the concept of the whole ecosystem. Special emphasis is given to the modelling of phenological events in their biological and abiotic environmental perspective. The key methods of modern phenological work are presented. These include observation networks, phenological gardens, *remote sensing* and mapping techniques. "Nature and Resources" Unesco, 1975, vol. XI, no. 3, p. 38.

Lagoons and estuaries as features of coastal zones

The importance of lagoons and estuaries as regional features of the edge of the sea, as the marginal sea-land interface, was pointed out by K. O. Emery:

"Regional aspects of estuaries and lagoons. Estuaries and lagoons with their marsh or mud-flat fillings constitute a much higher percentage of the world's coasts than is generally recognized. In fact, many of the largest cities have been built on the marshes or on "made" ground adjacent to them. Around the United States 80 to 90 percent of the Atlantic and Gulf of Mexico Coasts and 10 to 20 percent of the Pacific Coast consist of estuaries and lagoons." K. O. Emery, Am. Ass. Adv. Sci., Publ. no. 83, p. 9.

The coastal zones are the most important areas of our earth. They are only small in area compared with the 71 percent of the earth's surface which is covered by ocean water. However, much of the ocean's value, social as well as natural, is concentrated in this relatively small area – the "coastal zone". This term is becoming an all-inclusive one, encompassing the adjacent uplands, the shore, and the waters of the continental shelf. Lagoons, estuaries and bays are features of the coastal zone in the sense of the margin of the oceans as the triple interface sea – land – air. The marine and land belts of the coastal environment are in close relationship and dynamic interaction with each other and have an unique character.

Lagoons

In this publication most of the presented examples concern lagoons. Lagoons represent coastal resources and are becoming more and more important for human utilization: as tourist- and recreation-areas, as natural salt- and brackish-water-basins connected with the sea for fish- and algae-breeding (aquaculture) and as waterfowl breeding and living areas.

Especially because of the growing need of protein by the world population these features will probably become even more important in the future as industrial and harbour development areas thus rendering them vulnerable to oil and petroleum spills and degradation. Both for use as recreation areas and for fish- and algae-breeding (protein-production) areas, there is a vital need to keep the water in the lagoons clean, i. e. to protect coastal and lagoonal waters against pollution from land-based sources. This calls for a knowledge of water mass exchange and morphological processes. Lagoons are enclosed environments to which Unesco's "Convention of the World cultural and natural heritage" is applicable.

Along many coasts of the earth there are water areas of the sea which are partly or completely separated from it by accumulation forms, i. e. by the barrier islands or peninsulas. "Lagoon" as a collective term is intended to imply: lagoon (English), Lagune, Haff – Strandsee – Noor (German), etang, lagune (French), bai (Dutch), estero, laguna (Spanish), laguner (Swedish), liman (Slav).

A scientific monography on lagoons and lagoonal forms and processes is given by H. G. Gierloff-Emden, 1961, Nehrungen und Lagunen: Petermanns Geographische Mitteilungen, 2. Quartalsheft, Gotha.

Natural processes shape the physical form of the coastal zone – at least up to the time when man begins to alter it. Changes resulting from natural processes are relatively rapid along coasts, which are among the most dynamic of landscapes. This calls for frequent repetition of observations. Since lagoons exist and are enclosed by their barrier islands or peninsulas, it is noteworthy to point out that these islands or peninsulas are accumulation forms, caused by wave action, and longshore currents.

The barrier islands also serve as a physical barrier to protect the coast from severe erosion during coastal storms and hurricanes. Shallow bathymetry and frequent daily land and sea breezes combine to promote vigorous water circulation and strong turbulence, and to keep fine sediments in suspension.

Lagoons exist on tidal and non-tidal coasts. Tides not only affect water levels at the land – sea interface, but also in lagoons, estuaries and bays. Periodical wet and dry conditions of the inter-tidal zones occur as a result of tidal inundation on a diurnal or semidiurnal time scale. These phenomena and processes cause a structure of vertical zonation: 1) above High Water, 2) inter-tidal, 3) below Low Water. Sedimentation, alternating with erosion caused by tidal streams in tidal channels, effect a change in the features of lagoons.

The water balance of lagoons is usually a rather complicated system. Representing a tidal pumping system which can be called the "hydrodynamic respiration" of the lagoons (and estuaries), they are real transitory features between river discharge from the continent, inner lagoonal waters and saltwater from the sea with fast changing salinity of different water masses and mixture. High suspension of organic and inorganic material causes turbid water masses. In such an environment currents occur which alter dynamic environmental factors i. e.: river discharge; tidal stream channels (in the lagoon); longshore wave generation; waves and tides on the outer barriers (foreshore); ocean surface currents and deep ocean currents (offshore).

Estuaries

From the numerous definitions of the term "estuary" in the aspects of physiography, geology, hydrography, chemistry or biology, two have been selected:

"An estuary is a semi-enclosed coastal body of water which has a free connection with the open sea and within which sea water is measurably diluted with fresh water derived from land drainage."

Pritchard, D. W., 1967, What is an estuary: physical viewpoint. Pp. 3 in: G. H. Lauff (ed.), Estuaries, Am. Ass. Adv. Sci., Publ. no. 83, Washington.

And: "Estuaries are submerged features of the coastal zone developed by geological agents, influenced by physical and chemical processes, inhabited by organisms, and used by man. Understanding the dynamic attributes of estuaries leads to more effective management of them.

B. W. Nelson, 1972, Environmental framework of coastal plain estuaries. Geol. Soc. Am., Mem. 133, p. 1.

Nevertheless, there is no sharp separation between estuaries and lagoons.

"The definition of estuaries based on their physiography does not correspond to Pritchard, 1952: "An estuary is a semi-enclosed coastal body of water having a free connection with the open sea and containing a measurable quantity of seawater." This description could also apply to most of the marine lagoons, and explains the confusion often found in American biological publications on the subject of "estuarine waters" and "estuarine communities"."

H. Caspers, 1967, Estuaries: analysis of definition and biological considerations. Am. Ass. Adv. Sci., Publ. no. 83, pp. 6–8.

Lauff, G. H. (ed.), 1967, Estuaries. Am. Ass. Adv. Sci., Publ. no. 83, 757 p.

Pritchard, D. W., 1952, Estuarine hydrography. Adv. Geoph., vol. 1, pp. 243–280.

Stommel, H. M. and H. G. Farmer, 1952, On the nature of estuarine circulation. Woods Hole Oceanog. Inst., Ref. no. 52–88, 52–51, 52–63, Woods Hole, Mass.

Lagoons, estuaries and bays as biological ecosystems

The importance of lagoons and estuaries (and bays in some sense) which are very rich areas of fishes, mussels and crabs and are areas of intensive commercial and sport fisheries, has been pointed out frequently. For instance the total catch of estuarian and lagoonal species on the Gulf Coast of the USA has been estimated (in 1967) to be about 1.5 billion pounds, the catch of sport fishery to be more than 40 000 000 pounds. But very little is known of the ecological systems of these features or the life history of animals in marine-estuarian and marine-lagoonal cycles. So, any scientific method or tool which can help to analyse and understand lagoons, estuaries and bays, such as physiographical (geological, hydrographic, chemical) features and biological sites of our encironment should be used to reach a better understanding and, if possible, handling of our environment.

In lagoons, estuaries and bays there exist several biological features: light-using phytoplankton, zooplankton, zooplankton-eating fish, top layer carnivores, grazing zooplankton and fish on the bottom and waste decomposing micro-organisms at the bottom, together with complicated cycles of pathways of energy, minerals and food.

Some of these, e.g. the plankton, influence the colour of the water, as does suspended inorganic material, and produce water masses with particular characteristics. Lagoons and estuaries are intermediate filter-systems. Moreover there exist short and long periodical and biological cycles in the *time scales* (units) of tides or seasons and regional differences, depending on climatic, geological and hydrographic features. Exposure time and repetition rate usually is not in coincidence with these cycles. Using satellite imagery, it is therefore difficult, to select information of a certain category, such as water masses or bottom topography, causing such biological phenomena.

In this publication a number of regional studies are presented as case studies on remote sensing of the environment concerning lagoons, estuaries and bays.

Lagoons, estuaries and bays and remote sensing of environment

Remote sensing not only offers new methods of updating our knowledge of the coastal zones, but in addition provides new data, especially to the difficult problem of transport of suspended sediment from the lagoons to the sea. Keeping the lagoonal waters clean from pollution has become an important research project, due to increasing industrialization and use by man of these areas. Orbital remote sensing is a tool by which the hydrodynamic conditions of the water body exchanges can be studied. The objectives of this nearshore remote sensing research may lead to a better understanding of the problems being an attempt: to describe surface and near-surface fluid flow dynamics in a lagoonal or estuarine environment using synoptic orbital photography and imagery, to establish the temporal characteristics of various surface and near-surface coastal features, and to enhance the spatial and temporal characteristics of such a system.

F. B. Phleger, 1967, Some general features of coastal lagoons: Lagunas Costeras, Symposium, UNAM Unesco, p. 5–20.

Barrett, E. C. and L. F. Curtis, 1976, Introduction to environmental remote sensing: Chapman & Hall Ltd, London, 288 p. 165 fig.

Cronin, L. E., 1971, The role of Man in Estuary Process, pp. 266–294 in Th. R. Detwyler "Man's Impact on Environment"

Dolan, R., B. Hayden and J. Heywood, 1976, Managing coastal biome interfaces: Nature and Resources, Bull. of the Man and the Biosphere Progr., vol. 12, no. 1, Jan.–March, Unesco.

Emery, K. O. and R. E. Stevenson 1957, Estuaries and Lagoons: *Geol. Soc. Amer.* Mem. 67.

Gunter, G., 1967, Some relationship of estuaries to the fisheries of the Gulf of Mexico. *Am. Ass. Adv. Sci.*, Publ. no. 83, p. 621–638.

Lauff, G. H. (ed.), 1967, Estuaries: *Am. Assoc. Adv. Sci.*, Publ. no. 83, Washington.

Nelson, B. W. (ed.), 1972, Environmental framework of coastal plain estuaries. Symposium on estuaries, 18th meeting of southeastern section, The Geological Society of America, South Carolina, 1969: *Geol. Soc. Am.*, Mem. 133.

Odum, E. P., 1959, Fundamentals of ecology. 2d ed. Philadelphia and London: W. B. Saunders. 546 pp.

Odum, H. T., 1972, Functional classification of coastal ecological systems of the United States. *Geol. Soc. Am.*, Mem. 133, p. 9–28.

Odum, H. T., B. J. Copeland and E. A. McMahon (eds.), 1969, Coastal ecological systems of the United States: Federal Water Pollution Control Administration, Final Rep., Contr. RFP 68–128, 3 v., 1878 p.

Perkins, E. J., 1974, The biology of estuaries and coastal waters: Academic Press, London and New York, 640 p.

Unesco, IAHS, 1975, Hydrology of marsh-ridden areas: Studies and Reports in Hydrology, 19, 562 p., 100 fig. (resumes in French).

SKYLAB Earth Resources Experiments

"The *Skylab Earth Resources Experiment Package* (EREP), at a nominal altitude of 235 nautical miles (435 kilometers), used visible-light and near-infrared photography and infrared spectrography, an electro-mechanical scanner, and sensors for microwave surveys. Film and tape recording equipment gathered masses of remotely sensed data that will be used to establish the feasibility and usefulness of such earth-survey techniques.

The first Skylab crew was launched from the Kennedy Space Center on May 25, 1973 and completed 404 orbits of the earth at a nominal altitude of 235 nautical miles (435 kilometres) before returning on June 22. During 11 earth resources passes, the crew obtained 5,275 frames of imagery and 45,000 feet (13 716 metres) of magnetic tape data. The second crew was launched July 28, 1973 and completed 858 orbits that included 44 earth resources passes. The data, comprising 13 429 frames and 93 600 feet (28 529 metres) of tape, were returned to earth with the astronauts on September 25. The third crew, which took off on November 16, completed 1214 orbits and brought back 17,000 frames and 100 000 feet (30 480 metres) of taped earth resources data on February 8, 1974."

SKYLAB Earth Resources Data Catalogue, NASA, 1974, pp. VI–VII

EREP was designed to be a logical step in the application of equipment and techniques used in aerial surveys and the Earth Resources Technology Satellite (ERTS).

Earth Terrain Camera

This single-lens camera assembly had an f/4 lens with a focal length of 18 inches (45.7 centimetres) and was compensated for Skylab's forward motion. It was aligned within 1.4 degrees of the multispectral camera. The field of view of 14.24 degrees provided a ground coverage of about 59 nautical miles (109 kilometres) square. Film width was 5 inches (12.7 centimetres), which provides an usable image that is 4.5 inches (11.4 centimetres) square. Shutter speeds were 1/100, 1/140, and 1/200 of a second. Sequence photography intervals were possible from 0 to 25 frames per minute. Stereoscopic viewing with overlaps is not of importance for flat coastal areas.

The NASA-APOLLO 9 Multispectral Photography

The NASA Experiment S 0-65 multispectral photography flown on the APOLLO 9 mission, 8–12 March 1969 provides the first simultaneous satellite photography of the earth's surface in three distinct spectral bands spanning the visible and near-infrared regions of the spectrum taken by four cameras, each of which contained a different film-filter combination. The photography consists of four sets of photographs of identical surface locations, taken simultaneously.

Electrically driven Hasselblad cameras, model 500 EL, each equipped with a f/2.8 Zeiss 80-millimetre Planar lens, were used for Experiment S 0-65. The images of surface objects appear in the same co-ordinate positions on all four photographs in the multispectral set.

The set of four multispectral photographs is listed in the following table (from "Apollo 9 multispectral Photographic Information"):

Band (NASA designation)	Film and filter	Mean wavelength of sensitivity	Nominal bandpass
A	Infrared Ektachrome, colour SO-180, Photar 15	Green, red, and infrared	Total sensitivity of all dye layers, 510 to 900 millimicrons
B	Panchromatic-X, (black and white) type 3400, Photar 58	525 millimicrons, green	460 to 610 millimicrons
C	Infrared Aerographic (black and white) type SO-246, Photar 89 B	800 millimicrons, infrared	700 to 900 millimicrons
D	Panchromatic-X, (black and white) type 3400, Photar 25A	645 millimicrons, red	580 to 700 millimicrons

"The 400- to 900-millimicron range of wavelengths embraces all colours of the visible spectrum (violet, indigo, blue, green, yellow, orange, and red) together with wavelengths slightly longer than the visible red region of the spectrum, known as the near-infrared region. The 400- to 900-millimicron range is composed of four wavelength bands, of which three are in the visible region and one is in the infrared region of the spectrum. The three visible bands correspond to the three primary colours: blue, green, and red. An aerial or space photograph can be taken in any one of these bands by (1) using a photographic film sufficiently sensitive to energy in that band and (2) using a filter which, while transmitting energy within that band, effectively excludes energy of all other wavelengths to which the film is sensitive."

R. N. Colwell, 1970, in NASA Technical Memorandum X-1957.

Optimum wavelength for photographing natural phenomenon is for acquisition of maximum underwater detail the green and red band.

The blue band was not used in the S 0-65 Experiment owing to a lack of sharpness, due to atmospheric scattering.

In many instances, two types of objects may contain essentially the same reflectivity in one band, but different reflectivities in another band; that is, each type of object tends to exhibit an unique "tone signature" in multispectral photography. This tone signature is of great value in identifying the object.

Although a sequential or comparative interpretation of each of the four photographs comprising a multispectral set can yield significant information, a substantially more productive analysis can be made by simultaneous additive colour viewing.

The major advantage of multispectral photography is realized when the reflectivity of an object can be compared simultaneously in two or more portions of the spectrum. For specific application of NASA multiband photography compare:

APOLLO 9 Multispectral Photographic Information

NASA-Technical Memorandum NASA TM X-1957 April 1970

Compiled by J. L. Kaltenbach,

R. N. Colwell, D. Lowman, and E. Yost and S. Wenderoth

Manned Spacecraft Center Houston, Texas

National Aeronautics and Space Administration Washington, D. C.

Colwell, R. N., et al.: An Evaluation of Earth Resources using Apollo 9 Photography. (*Final report by the Forestry Remote Sensing Laboratory, School of Forestry and Conservation*, Univ. of California, under NASA contract NAS 9-9348), Sept. 30, 1968.

NASA–ERTS–1 satellite MSS imagery

ERTS-1-MSS images have been used by the author for thematic scientific and cartographic analysis. These images have been taken by an automatic sensor system.
In July 1972 NASA launched the first satellite of the ERTS (Earth Resources Technology Satellite since 1975 renamed "Landsat") series from Vandenberg, California. It circles the earth in 103 minutes at an altitude of 494 nautical miles (915 km with a tolerance between 912–920 km). Its quasi polar orbit is sun-synchronized. The satellite completes nearly 14 orbits per day. The successive orbit lies a bit more westerly because of the eastward rotation of the earth. Thus the earth can be observed between 81° N and 81° S latitude. Every point of the earth is reviewed by the sensor during 251 rotations which are completed in 18 days.

The distance apart of the (non-successive equator crossings) image strips is 1.43°, which equals 159 km at the equator. The sub-satellite track dissects the equator at an angle of 99.114°. The sub-satellite point (point perpendicular below the satellite on the earth's surface) moves with a speed of about 6 km/sec, which means it covers an image area (called "scene") in 30 sec. Reception time (data acquisition time) over the equator is always roughly 9.30 local time. The following day a neighbouring image strip overlaps the previous strip of an area, e. g., two adjacent images (called: "scene") are not synoptic. About 9 000 pictures are taken in the course of an 18-day cycle.
The conventional method of aerial photogrammetry is not replaced but supplemented by ERTS-MSS-images. The ERTS-satellite delivers exact vertical images. Due to the satellite's altitude, the high angular resolution of the sensor, and the small earth section recorded, each single image is practically orthographic, i. e. every point is shown as if the viewer were directly above it. Thus, in this case, the picture is practically an orthographic one.
With these scenes an area of about 100 nautical miles square (185 km x 185 km) is covered. Considering the scale and the size of the image scene, the curvature of the earth can be disregarded. Also, the relief distortions with respect to the projection surface can be neglected. Therefore topographic information, on smaller scales (< 1 : 250 000), can be mapped directly from the image. According to NASA, ERTS-1-images have the following maximal positioning errors:
Due to the curvature of the earth: 200 m; due to the deviation of 1° from the vertical: 440 m; due to relief differences in 1000 m heights: 160 m; due to atmospheric ray diffraction: 0.3 m. Together a maximum of 800 m, (after compensating the photogrammetric-geodetic error of 300 m, which equals about 0.1 mm on a scale of the 1 : 1 million map (International World Map) i. e. the error value is less than that given by paper distortion and drawing inaccuracy at this scale.

The Multispectral-Scanner-System (MSS) of ERTS-1-Satellite

The functioning sensing system consists of a Multispectral-Scanner-System (MSS) which in principal is a line-scanner. Multispectral scanners register different spectral bands out of the sunlight energy intensities as they are reflected from different objects of the scene. The energy recorded is dissected by the sensor into spectral bands: for each resolution element and spectral band, the measured energy intensity is transformed by the detector into an analagous electrical signal which is intensified and either transmitted in real time to earth or tape recorded in the satellite itself. The analogous data can be recorded on photographic film. The result is a photo-like image whose various grey tones represent different object reflectivities.
From the view of information theory in this case, the "photo" scene (images) can be considered to be the sender. It must be taken into account that, for example, ERTS-satellite images as sources of information also represent the end product of a complex communication system. The remote sensor from its platform, the satellite, measures signals which have come in over a complex transmission mechanism (radiation, atmosphere, waterbody-penetration). These signals are transmitted to the receiving station on the ground where they are decoded. During the whole process of recording and transmitting, the images which are later used as the source of information, are systematically influenced by factors inherent in the system.
The ERTS-satellite is equipped with a 4-channel multispectral scanner (MSS) which records earth's surface squares of 100 nautical miles square. The images are sent to earth in digital form, where the signals are processed into negatives on 70 mm film from which black and white paper copies or films are produced. The original MSS image is created as a photographic film from an electron beam recorder at a scale of

1 : 3.3 million (image size 2.2 inch). The scale of the final enlarged 7.3 inch square images is then 1 : 1 million, i.e. 18.5 cm × 18.5 cm, not exactly a geometric square, but a parallelogram, caused by the angle of orbit against the polar axis. The quality and resolution of ERTS-images are even better than was originally expected, so that new possibilities are offered for topographic and thematic mapping, e.g. for hydrology, oceanography and geology.
Due to the line-scanner system, the ERTS-image is built up from approximately 3 400 lines per image; these are visible on the image and show up clearly on in an enlargement of the image. Therefore the enlargement possibilities are limited. The lines are split up each into about 3 300 particular "points" of the scanner spot size (i.e. instantaneous field of view), called pixels each of which corresponds to an area of 79 × 79 m on the earth. These pixels are the data cells, "pixel" is the abbrevation for "picture element". The lines appear as line-structures which are non-existent in reality in the scene on the ground. This must be taken into consideration during interpretation.
Because of these scanning lines, MSS-images show no texture as is usual on photos taken optically by photogrammetric cameras with lenses. The only information carrier is the grey-tone and serves for interpretation. Therefore in many cases the help of a densitometer, which detects grey-tones more exactly than the eye, is necessary for optimal interpretation.
There is one important advantage of the scanner technique over normal photographs: image-to-image correlation according to the density value is directly possible because the scanner is recording continuously. The intensity of illumination of the image hardly varies, owing to the quick movement of the satellite relative to the clouds. The congruence of the four MSS-images is advantageous when comparing the representation of a scene in the four bands, as in the case of the colour composites.

The spectral bands of the MSS-images

The term "*channel*" is currently in use. However, since in this publication, the term channel has been employed in the sense of a "tidal channel" of the coastal environment, the term "*band*" has been used for the MSS-spectral sensors.
The MSS records in four channels of spectral bands. The division into four bands has the advantage over a panchromatic emulsion-sheet photograph of representing the object reflectivity in four spectral bands quantitatively. Also improved additional information can be obtained through the grey-tones with reference to the qualitative categories of the objects. Comparing images from different spectral bands assists with the identification of certain objects on the earth surface, it is possible to respond to and to recognize many phenomena by comparing two or more spectral bands. The individual pictures of the 4 different spectral bands can also be superimposed.
Colour composites can be produced (false colour pictures) by adding the grey tones to corresponding colour tones. (See cover of this book and example El Salvador, ERTS-1-MSS image).
During this process three of the four spectral bands are coded blue, green and red respectively; usually the green band of the spectrum (0.5–0.6 μm) is coded blue, the red band (0.6–0.7 μm) green, and the near infrared band (0.7–0.8 μm) red. Thus water appears deep blue due to its high reflection of green light; trees and plants (on the other hand) are shown red due to their very high reflection of the infrared spectrum by chlorophyll. Thus, the contrast water versus vegetation (blue to red) is obviously better than in nature and in the visible spectrum – the blue colour of water to green vegetation. Material such as granite appears grey and basalt lava black due to their non-selective absorption between 0.5 and 0.8 μm. These pictures are mostly used in oceanography and limnology in determining changes in water level, temperature and opacity as well as to detect pollution, suspended sediments in water (turbid water masses) and vegetation or glaciation.

Properties of the MSS-spectral bands

For the visible and "near infrared" bands of the spectrum (0.4 to 0.7 μm) one can state that:
Geological objects: splitting up into individual spectral-bands provides little new information because the different rock-materials in the area covered by this publication do not show well known specific spectral signatures:
Two medium imagery: submarine topography can be identified in the band 0.48 to 0.65 μm owing to penetration through the water;
Vegetation: the band 0.4 to 0.65 μm provides additional information, because chlorophyll (at 0.4 and 0.65 μm) shows distinct absorption maxima (e.g. mangrove vegetation on flats).

Satellite photographs taken during Apollo missions show that at 0.7 μm there is a very high rise of reflection from vegetation which appears in red on the so-called infrared false colour emulsion sheet films.
For inter-tidal zones: objects, due to their substratum, have different water content (sand bars and mud flats). "Sand" substratum dries out more quickly and warms up more quickly in the sun than mud. This influences the red band image (0.6 to 0.7 μm) so that it shows more and different contrasts of soil substratum, thus providing additional information.
The MSS image taken in the part 0.7 μm to 1.1 μm, (corresponding to bands 6 and 7), i.e. the reflective near infrared, show:
higher resolution, more contrasts, few grey tones, light-dark reversal of vegetation compared with sand-mud flats.
The choice of the spectral bands should guarantee an optimum of information for: structural and qualitative differentiation, (at least 10 densitometric determinable grey tones should be available); and identification of objects already detected (this is, for example, not assured in thermal infrared-detection of rock types).
Compared with direct observation by the human eye, there are significant differences in recording optical signals in individual spectral regions. The highest spectral sensitivity of the eye is between 500 and 650 μm, which is between green and orange in the visible light spectrum with a pronounced maximum at 555 μm.
The sensitivity for penetration through water of the human eye and panchromatic colour photograph, for certain phenomena, cannot be surpassed by any band of a multi-spectral scanner, but the MSS delivers better differentiation.

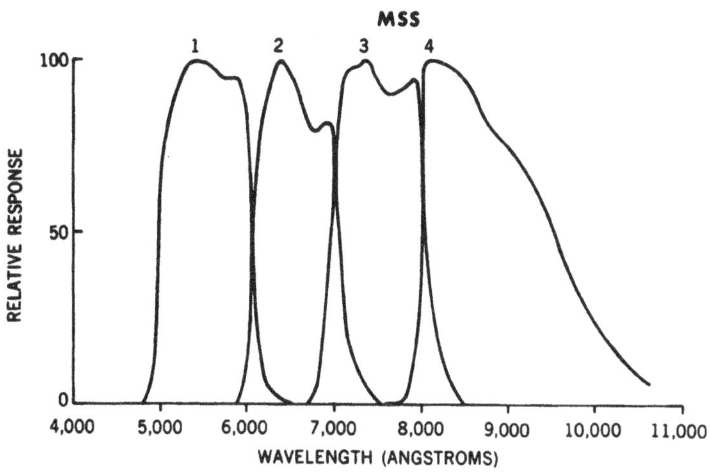

Spectral response for ERTS-I-MSS (Multispectral scanner)
Band 1, 2, 3 and 4 (channels) of MSS-ERTS-A-Satellite -NASA
(after ERTS Data Users Handbook, NASA).
1 Angström A = 10^{-8} cm.
μm = Micrometer, 1 μm = 10 000 Angström.
1 μ = 1 Micron = 10^{-4} cm = $^1/_{10000}$ cm.

Parameters dependent on objects

The surface reflectivity of the substratum in the four different spectral bands is recorded by the four channels of the ERTS multispectral scanner system, (this applies only when the sun's elevation is over 30°). (The visible band of the spectrum ranges from 0.4 to 0.7 μm).

Objects	Channel 4 band 0.5–0.6 μm (UV from 0,36 · 10^{-4} cm) green	Channel 5 band 0.6–0.7 μm orange	Channel 6 band 0.7–0.8 μm red	Channel 7 band 0.8–1.1 μm (IR from 0.78 · 10^{-4} cm i. e. reflective, non thermal infrared
Water	*3.75*	*2.24*	*1.20*	*1.89*
Sand	*5.19*	*4.23*	*3.46*	*6.71*
Ice	*18.30*	*16.10*	*12.20*	*11.00*
Snow	*19.10*	*15.00*	*10.9*	*9.2*

Properties for recording of objects and for analysis:

	channel 4:	channel 5:	channel 6:	channel 7:
	low contrast	*moderate contrast*	*stronger contrast*	*stronger contrast*
	good penetration through water (transparency)	*best penetration through water (transparency)*	*very poor transparency*	*no penetration through water*
	low contrast	*good contrast*	*good contrast*	*very good contrast*
	vegetation and soil not separable	*vegetation against soil*	*for various soils grey tone according to soil water content*	*urban settlements, soil-vegetation grey-tone reversal in comparison to red band: vegetation bright*
	least resolution	*moderate to good resolution*	*highest resolution*	*good resolution*

General characteristics can differ in individual cases, due to object dependent parameters (absolute water depth, scattering at sediment clouds in water according to size of particle).
Water and ice contrast is large in all bands.
Water and sand contrast is largest in the infrared bands.
The absorption of water is greatest in the 0.8 to 1.1 μm band, i. e. water appears black. Water penetration is three times as good in the green band than in the blue band; it can be best for shallow water in the orange band when suspended material is present in the water. Spectral bands which penetrate water and record submarine topography cannot show any sharp coastal border lines (water lines), i. e. terrestrial beach against submarine foreshore.
Infrared gives good penetration of haze in the atmosphere. Vegetation in channel 4–6 has a lower reflectivity due to the strong absorption of chlorophyll, vegetation on channel 7 (IR) has strong reflectivity. This applies to the example of mangrove given in this publication.

Data and Catalogues

Obtaining data from NASA's earth resources observation systems
(orbital photography and imagery)
(as advertised in NASA Skylab Earth Resources Data Catalogue, 1974)

"In accordance with the National Aeronautics and Space Act of 1958 and the Freedom of Information Act of 1967, NASA provides earth resources data to three federal data outlets. These provide the data to the general public at a nominal charge."
In 1972, the US Secretary of Agriculture designated the *Aerial Photography Field Office* to receive and store all satellite imagery for the Department of Agriculture and the general public. This function is performed by:
Western Aerial Photography Laboratory, Administrative Services Division
ASCS/USDA 2505 Parley's Way. Salt Lake City, Utah 84109
Black-and-white and colour reproductions of satellite imagery can be ordered from the laboratory.

The *US Department of the Interior* established the *Earth Resources Observation Systems* (EROS) *Data Center in Sioux Falls*, South Dakota, in 1971. The Data Center is operated for the Department's Earth Resources Observation Systems Program by the Topographic Division of the Geological Survey to provide the general public, U. S. government agencies, and foreign governments with access to satellite imagery, NASA aircraft remote-sensor data, and U. S. Geological Survey photographs. The Data Center has facilities for data storage, retrieval, reproduction, and dissemination, as well as user assistance and training. For service from the EROS Data Center, contact:
U. S. Department of the Interior, EROS Data Center User Services, Sioux Falls, South Dakota 57198

The *National Oceanic and Atmospheric Administration* (NOAA), through the National Environmental Satellite Service, has established a data center in Asheville, North Carolina. For Service from this center, contact*:

Director, National Climatic Center Federal Building
Asheville, North Carolina 28801

* NOAA personnel should contact the NOAA *Earth Resources Data Center*, National Environmental Satellite Service, S 1162, Room 1069, Federal Office Building 4, Washington. DC 20233

For Gemini and Apollo photographs collection contact:
Technology Application Center University of New Mexico. Albuquerque, New Mexico 87106
For Gemini, Apollo and Skylab photographs contact:
Earth Resources Division, NASA Manned Spacecraft Center (MSC), Houston, Texas.

Stevenson, R. E. and Nelson, M., "An Index of Ocean Features, Photographed from Gemini Spacecraft", Contribution no. 253, Bureau of Commercial Fisheries, Biological Laboratory, Galveston, Texas, prepared for NASA MSC.

Apollo 9 Plotting and Indexing Report
October 1969, Technical Report LEC/HASD no. 671-80-037
NASA Manned Spacecraft Center Mapping Sciences Laboratory
Prepared by Lockheed Electronics Company for Mapping Sciences Laboratory under NASA Contract NAS 9-5191 LEC Document 648D.21.037

Skylab earth resources data catalogue: National Aeronautics and Space Administration, Lyndon B. Johnson Space Center, Houston, Texas, 1974.
For sale by the Superintendent of Documents, U. S. Government Printing Office, Washington D. C. 20402
Stock No. 3300-00586
This book provides a complete index of EREP photographs and information on how all EREP data can be obtained. It also gives examples and explains potential uses of the more than 35,000 frames of earth resources imagery and 238,600 feet (72 725 metres) of taped earth resources data.

NASA-Skylab-Missions
Earth Resources Experiment Package (EREP)
S-190 B Photographs with Earth Terrain Camera (ETC)
The Earth Terrain Camera (ETC) utilizes 5 inch (127 mm) film supplied in cassettes of approximately 450 frames each. The camera is equipped with a f/4 lens with a focal length of 18 inches (0.4572 metres), providing ground coverage of approximately 59 nautical miles (109 kilometres) on each side.

Sensor Coverage Summary Map Description
The S-190B Sensor Coverage Summary Map contains the approximate coverage limits of S-190B photographs taken during the three manned Skylab missions, and the outlines and names of eight larger Scale (1 : 3 700 000 to 1 : 15 000 000) Sensor Coverage Index Maps that will be produced to index the principal point, frame number, and sufficient frame size data to establish the area of coverage for each S-190B photograph (from Skylab Earth Resources Catalog).

ERTS Data Users Handbook, 1972, NASA/GSFC Doc. no. 71SD4249, Washington D. C.

EREP Users Handbook, 1971, NASA, Manned Spacecraft Center, Houston, Texas.

ESRO, 1974, European Earth Resources Satellite Experiments: Proc. 1st Symp. on Europ. Earth Resourc. Satell. Exper. Frascati, ESRO Paris.

Eros Data Center Standard Products
Satellite Products Sept. 1, 1974

ERTS data

Image Size	Scale	Format
2.2 inch.	1 : 3 369 000	Film Positive
2.2 inch.	1 : 3 369 000	Film Negative
7.3 inch.	1 : 1 000 000	Film Positive
7.3 inch.	1 : 1 000 000	Film Negative
7.3 inch.	1 : 1 000 000	Paper
14.6 inch.	1 : 500 000	Paper
29.2 inch.	1 : 250 000	Paper

Colour Composite Generation* (When not already available)

Image Size	Scale	Format
7.3 inch.	1 : 1 000 000	Printing Master

* Colour composites are portrayed in false colour (infrared) and not true colour.

NASA ERTS Catalogs

Title
U. S. Standard Catalog – Monthly
Non-U. S. Standard Catalog – Monthly

Cumulative U. S. Standard Catalog – 1972/1973
Volume 1, Observation ID Listing
Volume 2, Co-ordinate Listing

Cumulative Non-U. S. Standard Catalog – 1972/1973
Volume 1 Observation ID Listing
Volume 2 Observation ID Listing
Volume 3 Co-ordinate Listing

SKYLAB Photography

S190A

Image Size	Scale	Format
2.2 inch.	1 : 2 850 000	Film Positive
2.2 inch.	1 : 2 850 000	Film Negative
6.4 inch.	1 : 1 000 000	Paper
12.8 inch.	1 : 500 000	Paper
25.6 inch.	1 : 250 000	Paper

S190B

Image Size	Scale	Format
4.5 inch.	1 : 950 000	Film Positive
4.5 inch.	1 : 950 000	Film Negative
4.5 inch.	1 : 950 000	Paper
8.6 inch.	1 : 500 000	Paper
17.2 inch.	1 : 250 000	Paper
34.4 inch.	1 : 125 000	Paper

Standards, definitions and terms

Photointerpretation, remote sensing

Fachwörterbuch, Benennungen und Definitionen im deutschen Vermessungswesen. Technical dictionary, terms and definitions as used in surveying and mapping in Germany: vol: 7, photogrammetry, photo-interpretation: Vorläufige Ausgabe zum XIII. Internationalen Kongreß der F. I. G. Wiesbaden 1971: Verlag des Institutes für angewandte Geodäsie, Frankfurt a. M. 1971.

NASA, Data users handbook, Earth Resources Technology Satellite, GSFC Document 71504249, 15. September 1971, revised 15 February, 4 May, 18 July, 8 September, 15 September and 17 November 1972: Goddard Space Flight Center, Greenbelt.

Allen, W. H., 1965, Dictionary of technical terms for aerospace use: NASA 5, p. 7.

Reeves, R. G., 1976, Glossary: in Manual of remote sensing, vol. 2, *Am. Soc. Photogramm.*, p. 2061–2110.

Oceanography and coastal geomorphology

Symbols and abbreviations used on Admiralty Charts: Chart 5011, book edition 2 – November 1973, published at the Hydrographic Department, Taunton, under the Superintendence of Rear-Admiral G P D Hall, D. S. C., Hydrographer of the Navy.

Baker, B. B. jr., W. R. Deebel and R. D. Geisenderfer (eds.), 1966, Glossary of oceanographic terms, 2nd edition: U. S. Naval Oceanographic Office, Spec. Publ. 35, Washington D. C.

Hunt, L. M. and D. G. Groves, (ed.) 1965, A glossary of ocean science and undersea technology terms. An authoritative compilation of over 3 500 engineering and scientific terms used in the field of underwater sound, oceanography, marine sciences, underwater physiology and ocean engineering: Compass Publications, Inc., Arlington, Virginia.

Fairbridge, R. W. (ed.), 1966, The encyclopedia of oceanography: Encyclopedia of earth sciences series, vol. 1, Reinhold Publishing Corporation New York.

Glossary of technical terms in cartography, 1966: The Royal Society, Sub-commission of the British Nat. Comm. Geogr., London, 84 p.

Geography, geology

Meynen, E., 1974, Bibliography of mono- and multilingual dictionaries and glossaries of technical terms used in geography as well as in related natural and social sciences: Intern. Geogr. Union, Comm: Intern. Geogr. Terminol., Franz Steiner Verlag, Wiesbaden.

Stamp, L. D. (ed.), 1961, A glossary of geographical terms, prepared by a committee of the British Association for the Advancement of Science: Longmans, London.

Dictionary of geological terms, prepared under the direction of the American Geological Institute. 7 500 items: Anchor Books, Anchor Press, Doubleday, Garden City, N. Y. 1974.

A multiple boundary layer model of air — sea interaction (sand-storm, currents, turbidity and submarine topography) as a problem of satellite photography

Eastern Gulf of Venezuela, Caribbean Sea
Golfo de Coro, Peninsula de Paranagua

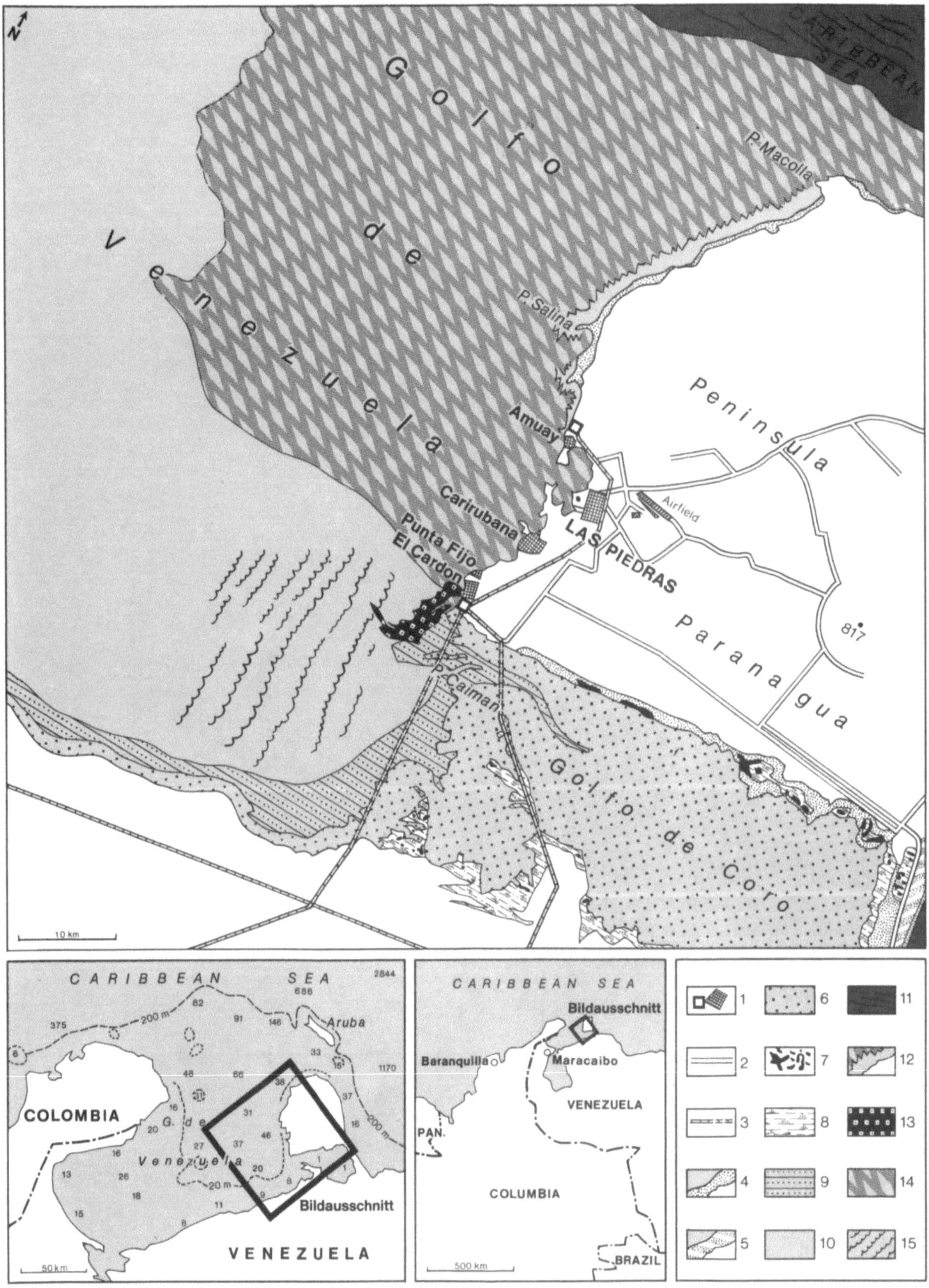

N
Golfo de Venezuela
CARIBBEAN SEA
P. Macolla
P. Salina
Peninsula Paranagua
Amuay
Carirubana
Punta Fijo
El Cardon
LAS PIEDRAS
Airfield
817
Caiman
Golfo de Coro
10 km
CARIBBEAN SEA
Aruba
200 m
COLOMBIA
G. de Venezuela
20 m
Bildausschnitt
VENEZUELA
50 km
CARIBBEAN SEA
Bildausschnitt
Baranquilla
Maracaibo
VENEZUELA
PAN.
COLUMBIA
BRAZIL
500 km
1
2
3
4
5
6
7
8
9
10
11
12
13
14
15

Eastern Gulf of Venezuela, Caribbean Sea
Golfo de Coro, Peninsula de Paranagua

Interpretation map, from satellite colour photograph

SKYLAB mission 3

Interpreter H. G. Gierloff-Emden, 1975

Legend of topographic and hydrographic features classified in zones with selected discriminated objects, (generalized: no differentiation of distinct water levels on this small-scale map).

I above High Water level: terrestrial

mainland on the map as white area

1 settlement and industrial area (refineries), towns and harbours

2 roads

3 pipeline (partly submarine in the Golfo de Coro)

4 coastal dunes

5 beach, sandy

II between High and Low Water level: inter-tidal

5 beach, partly inter-tidal, but not discernable on satellite photograph

6 tidal flats (in the Gulf of Coro not diurnal uncovered, compare III 6)

7 small coastal lagoons and salt pans and pans in marshes

8 marshes (partly in the Gulf of Coro)

III below Low Water level: marine

6 very flat bottom of the Gulf of Coro, less than 1 m deep, partly tidal flats (pictorial information gives no exact detection and identification)

9 foreshore

10 water of the Gulf of Venezuela

11 water of the Caribbean Sea,
to the channel north of peninsula Paranagua versus island Aruba; current pattern of the sea

12 surface of the sea of the gulf,
west of the peninsula compare 14

13 pollution from industrial area of El Cordon

IV atmospheric – oceanic phenomena, features of an ambiguous picture

(multiple boundary layer)

14 sand storm over the sea, covering sea-surface, waves (seastate), in rhomboidal and stripe pattern; pattern near coast smaller scale; pattern over outer gulf larger scale

15 clouds (atmospheric) or submarine sandwaves (oceanic-marine) features and pattern are detectable, but not clearly identifiable, created oceanic or atmospheric; pictorial information is ambiguous

Satellite photograph:

Eastern Gulf of Venezuela, Caribbean Sea
Golfo de Coro, Peninsula de Paranagua

Flight information data

Mission: *NASA – SKYLAB – 3*

Date: *18. Sept. 1973*

Photograph: *colour*
paper, image size 8.6 inch on scale 1 : 500 000

Magazine: *S – 73* *frame no. 35079*

Camera: *S 190 B, Earth terrain camera (ETC)*

Lens: *f: 4/450 mm*

Film: *120 x 120 mm*

Altitude above mean sea level: *234 naut. miles*

Geographical position of the principal point:
latitude: *11 degrees 45 min. N*
longitude: *70 degrees 30 min. W*

Chart: *ONC – WAC code K – 26*

Local observation time: *12.30*

Sun azimuth: *185°*

Sun elevation: *80°*

Photograph viewing mode: *vertical*

Cloud cover: *2 %, alto cumuli*

Scale of the printed picture: *1 : 455 000*

Ground coverage: *109 km x 109 km*

Ground resolution: *20 m to 80 m, depending on object and pattern*

NASA – SKYLAB 3

A multiple boundary layer model of air – sea interaction (sand-storm, currents, turbidity and submarine topography) as a problem of satellite photography

Eastern Gulf of Venezuela, Caribbean Sea
Golfo de Coro, Peninsula de Paranagua

NASA colour photograph; Skylab mission 3; S 190 B, S-73-35079; date Sept. 1973; flying altitude above mean sea level is 435 km; the scale of the photograph is approximately 1 : 455 000 (picture scale); ground resolution is 50 to 80 m; the photograph is approximately vertical. The photograph was taken about noon (local time) so that the azimuth of the sun is southerly; the sun's elevation is very steep, (approximately 80 degrees), the upper margin of the picture is orientated to the north-west.

This satellite photograph, taken from NASA's Skylab represents not only a spectacular colour space photograph but an unique synoptic momentum picture of the dynamics of oceanographic and meteorological processes, both, of the Golfo de Coro in the eastern part of the Venezuelan Gulf and the peninsula of Paranagua, the eastern frame of the Gulf.
This picture presents a variety of information on the following interaction between continental and oceanic phenomena and dynamics:

1. Sediment transport of blown sand in the atmosphere, i. e. a layer over the oceanic water, influencing the transparency of the atmosphere (atmospheric features),
2. reflexion from the oceanic surface, i. e. boundary layer (air – sea interface and wind – wave interaction),
3. sediment transport in oceanic waters, caused by currents and waves, i. e. layers of oceanic waters influencing the transparency of the waterbody,
4. submarine topography of the lowest boundary layer (sea – bottom interfaces) (this has a more static characteristic in comparison with phenomena 1, 2 and 3).

The interpreter is confronted with a complex model of phenomena and interaction processes; the photograph provides a structural and coloured pattern, but as ambiguous information. It is important to note that the tone of the colour of this picture depends on the developing process (copy generation), and in a book like this one additionally on the printing process. In this very case the water of the ocean appears blue, i. e. close to NASA photograph copy; there exist other copies of this photograph where the water of the ocean in general comes out in a more greenish colour.

Description of the area

Paranagua, a peninsula (60 km long, 50 km wide) surrounds the small Gulf of Coro, and represents the eastern frame of the Gulf of Venezuela with Maracaibo inlet. It is connected by the small flat barrier island of the Isthmo de los Médanos, which is 30 km long and 5 km wide and studded with dunes ("médanos") 15 m high (lower right corner of photograph). The peninsula of Paranagua is in general low and flat, but is 800 m high in its middle part. This is important as it affects the wind field. It has a core of geologically very old rock material causing the colour of the surface, where the bar rock is exposed.
North of the peninsula of Paranagua the island of Aruba is located at a distance of 30 km. This is a narrow channel, where the sea often becomes rough, when the winds blow through it.
The Gulf of Coro, whose only one metre deep floor appears grey-pink through the shallow water, is 40 km long and 30 km wide and nearly closed by a very small, long barrier island. It dries out partly at low water. The peninsula of Paranagua is partly surrounded by small coral reefs.
In 1500 and 1501, the coast was discovered and visited by Alonso de Ojeda, Juan de la Cosa and Amerigo Vespucci. At the landward origin of the barrier island de los Médanos, the settlement of Coro was founded in 1527, and is today the capital of the province of Falcon. Coro was the base for the conquests won by the "Welser", a trading firm in Augsburg, Bavaria. Because of the inhabitants' pile dwellings, the country was named "Little Venice" – Venezuela.
Oil production started in the region in 1876, when the first concessions were issued. In the thirties drilling had been carried out from artificial offshore platforms in the shallow gulfs. Since only small ships can pass through the

shallow entrance to the Lake of Maracaibo, pipelines were constructed to the peninsula of Paranagua, across the small barrier island which closes the Gulf of Coro, to Punta Fijo, Punta Cardon and Amuay, the ports for the export of the oil obtained from the Gulf of Maracaibo. Punta Fijo with its refineries is clearly distinguishable in the satellite photograph; the airport is situated nearby. Because of the oil industry in this area, detailed cartographic mapping was carried out as early as 1905. Pollution from industry at El Cardon spreads out as a plume of 10 km extension into the Gulf.

Hydrography and meteorology

Steady currents from a northern branch of the South Equatorial Current and from a southern branch of the North Equatorial Current flow from the east into the Caribbean Sea. With the southern marginal band which flows along the northern coast of Venezuela, and is called the "Caribbean Current", the average current velocity ranges to 1 knot (nautical mile per hour). Branches of this current pass the islands of Curaçao and Aruba and the north coast of the peninsula of Paranagua.

These currents have an unusual dynamic behaviour since the water masses have to enter through the narrow and shallow passages of Lesser Antilles, so that the current seems to oscillate, similar to tide phenomenon (as described by G. Schott, 1944).

In the Caribbean Sea, due to the North-east Trade Wind a current to the west is generated, so that an intensive translation of water masses along the coasts occurs: Pillsbury described the complex of the water mass translation in that area as: "Not a stream, but a sea is in motion", and P. Martyr reported in 1515, based on the expressions of the old Spaniard pilots:

„Man kann die ganze Karibische See mit ihren Strömungen fahrend von Osten nach Westen in 4–5 Tagen durchsegeln. Aber die Rückkehr von da ist infolge des konträren Laufes der Gewässer so mühselig und schwierig, daß es scheint, als ob die Schiffe einen hohen Berg hinaufsegeln und gegen die Gewalt Neptuns ankämpfen müssen." G. Schott, 1944, p. 276–277.

Winds in this area are dominant and prevailing from east with speeds of 5 to 6 Beaufort. There is an influence of orographical features of the peninsula on the winds (Katabatic winds). Squalls of short duration do occur and quickly roughen shallow waters, especially when blowing counter to the current and produce sediment in suspension over the flat sea bed. Dust clouds occur in the air. Dust sedimentation on the sea surface and in the upper layers of the sea causes discolouration of the water. A haze of salt-spray comes off the breakers and causes more blue light effect on the colour photography.

Phenomena, processes and interpretation

The tidal regime in this region is complex, being a mixture of diurnal and unequal semidiurnal types. The tidal range varies from 0.3 m to 1.5 m, average 0.6 m. At the moment when the photo was taken, there was an ebb-stream close to tidal Low Water. Water masses moved to the west, and the tidal flats, sandy and muddy, in the innermost part of the gulf seem to be dry but are covered by a very flat body of water, i. e. less than 1 m depth. (Compare grey-pink colour of the photo in the lower right).

Wind-blown sands from the peninsula dunes are transported as a real sand-storm to the west, and cover large areas of the gulf, so that a furrow of rhomboidal pattern (on a scale of 1 to 5 km in that network) of a grey-blue colour occurs and covers the ocean surface as a screen with some transparency. Other boundary layer phenomena, such as waves, plumes of sediment-laden turbid water or even sandwaves on the sea floor may be present, but these cannot be detected through this screen of wind-blown sand, so that discolouration of the water is ambiguous for the interpreter.

In the lower left part of the photograph it is uncertain whether the white-grey stripes, direction from N to S, are so-called giant sandwaves, i. e. submarine topographic features, they could well be an atmospheric feature, such as clouds. The pictorial information is ambiguous, and does not differentiate between patterns those created by clouds and those which are not.

In the upper part of the photograph, the ocean surface appears to have fine, long stripes; these may be caused by ocean currents. The pattern of these currents, as visible on the surface, may be formed due to the channel between the peninsula of Paranagua and Aruba island.

The visibility on the ground may be entirely different from the visibility from spacecraft because from the latter the view penetrates vertically (in a short way) through the layers. It is reported that haze and dust may deteriorate the ground visibility to as low as 6 km, and that the upper edge of the haze and dust cloud is sharply defined.

Sea-haze caused by salt-spray from breakers is present along the coasts and this also influences the visibility from spacecraft and the colour of the photograph in the coastal belt.

Sand-storm or dust-storm as a boundary layer over the surface of the ocean

Since still very little is known about these phenomena, some remarks should be given:
R. A. Bagnold, 1954, who wrote in his book: "The physics of blown sand and desert dunes", chapter 2, pages 10 and 11 on: The behaviour of sand grains in the air:

"The phenomenon of sand-driving
A popular misconception exists regarding sand storms, due to a failure to distinguish sand from dust. When, in any arid country, after a spell of calm weather, a strong wind begins to blow from a new direction, the air becomes charged with a mist of small particles. Where the surface is alluvial, with little or no sand on it, such as in Iraq or the country round Khartoum, the dust rises in dense clouds to a height of several thousand feet and the sun is obscured for a long period. This is obviously a dust storm, though it is often wrongly designated by the possibly more thrilling and cleanly term 'sand storm'. Owing to their feeble terminal velocity of fall the very small dust particles are raised and kept aloft by the upward currents of the wind's internal movement, . . ."
"Except in broken country, the sand cloud seems to glide steadily over the desert like a moving carpet, and the wind is comparatively gustless. When the wind drops the sand cloud disappears with it. This is a true sand storm."
"Even when the sand cloud is free from dust, as over a dry sand beach or in the interior of denuded desert, the actual movement of the sand within the cloud is very difficult to see. The individual grains are too small and travel too fast for the eye to follow their motion."

Remote sensing and change detection

The remarkable features of this satellite photograph are the phenomena:
sediment transport of blown sand in the air (atmospheric features) and
sediment transport of currents in oceanic waters (oceanic features)
Both are processes of small time scale, i. e. in unit time parts of one hour or several hours, and both features are of about the same metric scale (of unit length dimension). Shape and colour density of both phenomena are similiar. Therefore in this case change detection with spacecraft photograph series would be difficult; a system with photograph series of a time sequence of every 5 minutes would be necessary to reproduce the dynamics of these processes like a movie.
But the extension of these phenomena over a large area in metric dimensions of meso scale to large scale, i. e. some kilometres to 100 km, calls for an additional synoptic overview as provided by this space photograph.
It does not seem possible to obtain any further explanations from the author. But this satellite photograph represents one of the first views of the phenomenon of such a sand- or dust-storm as a boundary layer on oceanic water: The size, inner structure and scale of the rhomboidal mosaic of the sand carpet may help geophysicists to describe the process.

Bibliography

Arnold, J. E. and A. H. Thompson, 1969, Apparent coastal water discolorations from spacc, occanic and atmospheric: in Oceans from space, p. 82–91.

Bagnold, R. A., 1954, The physics of blown sand and desert dunes. London, 265 p.

Riehl, H., 1954, Tropical meteorology. London.

Schott, G., 1944, Geographie des Atlantischen Ozeans. 438 p., Hamburg.

Aviation meteorology of the West Indies, 1959. *Meteorol. Rep.* no. 22, pp. 85, Air Ministry, Meteorol. Office, London, Her Majesty's Stationery Office.

Westindien-Handbuch, 1958, I. Teil: Die Nordküste Süd- und Mittelamerikas. DHI, Hamburg, no. 2048.

Maps

Carta Geografia, Coro, in scale 1 : 250 000, Ministerio de Obras Publicas, Cartografiá Nacional, E.E.U.U. de Venezuela, no. 13.

Chart ONC – WAC Code K 26.

Nautical Charts: German: no. 443, no. 534; Dutch: no. 223; British: no. 1529, no. 1963; USA: H. O. no. 5437, no. 5520.

Transparency of clear oceanic waters and mapping of submarine topography, using satellite photography

Golfo de Guacanayabo, Caribbean Sea, south of Cuba

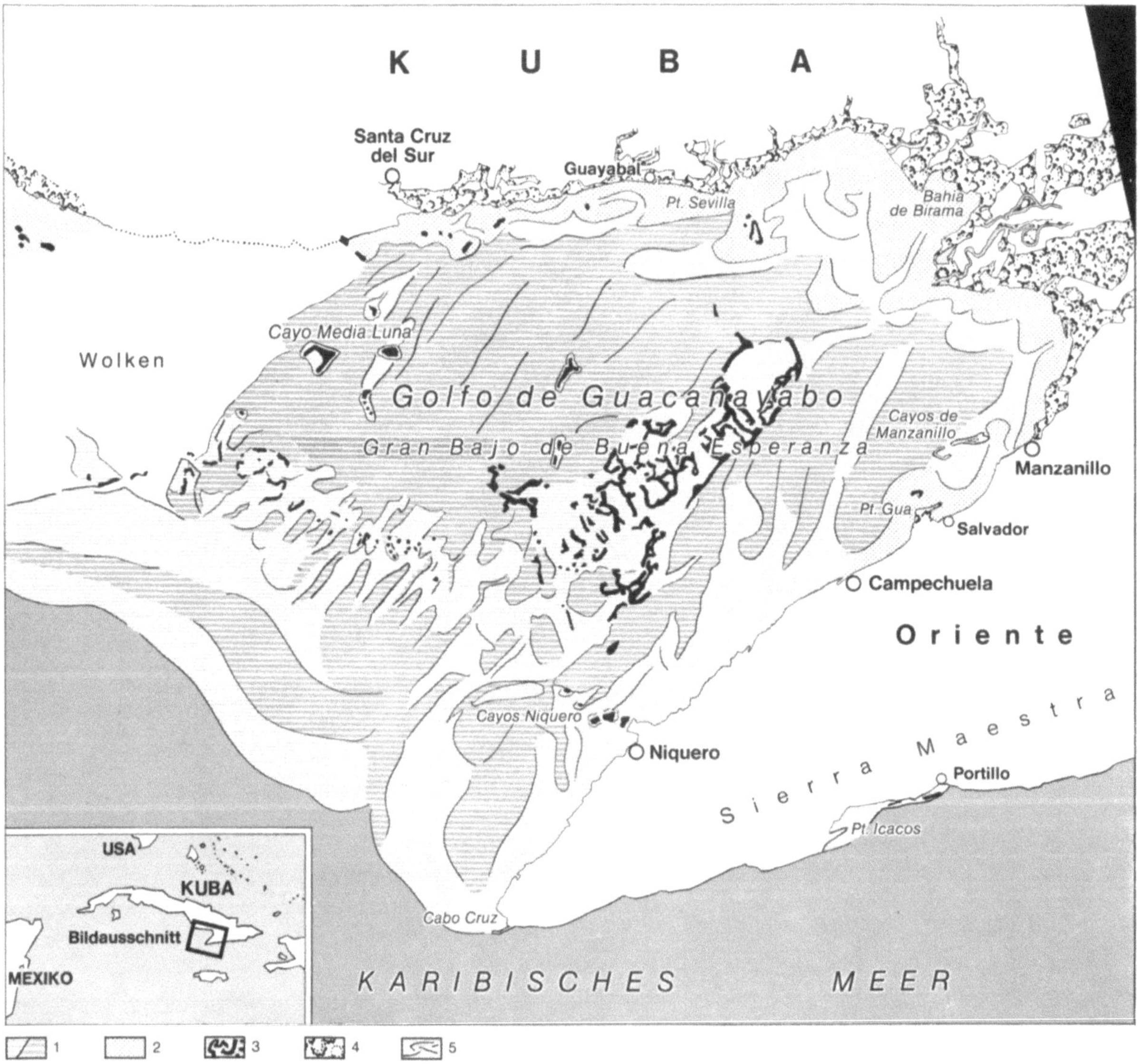
K U B A
Santa Cruz del Sur
Guayabal
Pt. Sevilla
Bahia de Birama
Cayo Media Luna
Wolken
Golfo de Guacanayabo
Gran Bajo de Buena Esperanza
Cayos de Manzanillo
Manzanillo
Pt. Gua
Salvador
Campechuela
Oriente
Sierra Maestra
Cayos Niquero
Niquero
Portillo
Pt. Icacos
Cabo Cruz
USA
KUBA
Bildausschnitt
MEXIKO
KARIBISCHES MEER
1
2
3
4
5

Golfo de Guacanayabo, Caribbean Sea, Cuba

Interpretation map, from NASA satellite colour photograph

APOLLO mission 9, no. AS 9-B-21-3314

Interpreter H. G. Gierloff-Emden, 1975

Legend of topographic and hydrographic features classified in zones with selected discriminated objects (generalized: no differentiation of distinct water levels on this small-scale map)
Scale approximately 1 : 1 000 000

I above High Water level: terrestrial

mainland on the map as white area, including Rio Cauto dry delta

3 foreshore coral reef and
coral reef (key) in the gulf
(key or "cay": low insular bank of coral, partly with sand, permanently dry and not covered)

II between High and Low Water level: inter-tidal zone

3 foreshore coral reef and coral reef (key) in the gulf, tidal covered (differentiation between I 3 and II 3 is not possible with satellite photograph only)

4 mangroves and inter-tidal flats

III below Low Water level: lagoonal and marine

3 foreshore coral reef and coral reef (key) in the gulf, permanently covered (differentiation between II 3 and III 3 is not possible with satellite photograph only)

1 bottom of the gulf, concerning the sea floor in the gulf within water depths mainly between 10 m and 20 m (compare particular chart with bathymetry) occurs in greenish-blue and blue colour on the printed satellite photograph

2 turbid waters over flat bottom of the foreshore (0 m to 10 m depth)

5 boundaries of water masses or indicating submarine topography of the bottom (concerning floor of the gulf), i. e. ambiguous information, depending on water penetration (transparency)

Caribbean Sea (on the map as "Karibisches Meer", i. e. oceanic water on deeper bottom (to a depth of 200 m) occurs on the printed satellite photograph in dark purplish-blue colour,
occurs on the NASA photographic copy more in purplish-blue colour,
contrast to III 5 is sharp, due to the steep continental slope oceanwards from shelf edge, which is close to 20 m contour line (compare particular chart with bathymetry)

Satellite photograph:

Golfo de Guacanayabo
Caribbean Sea, Cuba

Flight information data:

Mission: *NASA – APOLLO 9*

Date: *9. March 1969*

Photograph: *colour*

Magazine: *B* *frame no. 21–3314*

Camera: *Hasselblad 500 C*
70 mm photography

Lens:

Film: *70 mm Ektachrome*

Altitude above mean sea level: *102 naut. miles*

Geographical position of the principal point:
latitude: *20 degrees 28 min. N*
longitude: *77 degrees 38 min. W*

Chart: *ONC – WAC code Y – 26*

Local observation time: *14.30*

Sun azimuth: *200°*

Sun elevation: *47°*

Photograph viewing mode: *slightly oblique*
exposure axis to 20°

Cloud cover: *30 % alto cumulus*

Scale of the printed picture: *1 : 1 000 000 (approximate), varies. (Section of Satellite photograph)*

Colours of the printed picture (Munsell scale):
The region of the Gulf occurs more in greenish blue colour in comparison to the blue colour on the NASA copy.
The region of the Caribbean Sea occurs more in dark purplish blue colour in comparison to the purplish blue colour on the NASA copy.

NASA – APOLLO 9

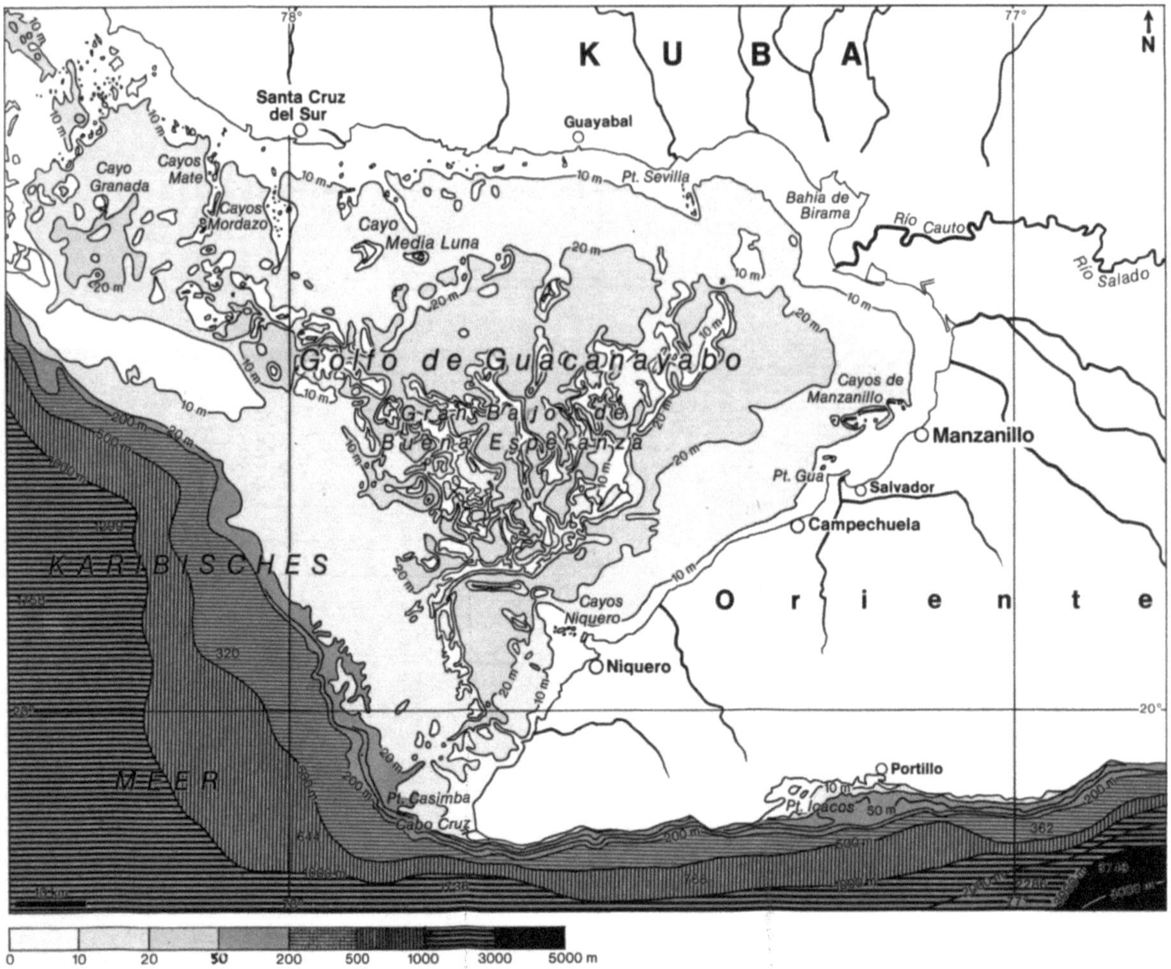

Golfo de Guacanayabo, Caribbean Sea, south of Cuba

Chart of the area of the satellite photograph
scale 1 : 1 000 000,
with bathymetric topography, depth in m

Transparency of clear oceanic waters and mapping of submarine topography, using satellite photography

Golfo de Guacanayabo, Caribbean Sea, south of Cuba

NASA colour photograph, Apollo mission 9, no. 21–3314; date 9. March 1969, "Cuba Oriente"; hand operated camera; 70 mm film; scale of this photograph approximately 1 : 1 000 000 (printed picture scale).
The photograph is not exactly vertical but is slightly oblique. It was taken in the early afternoon, about 15.00 local time, so that the sun's azimuth angle comes from SSW, i. e. with the same direction of the exposure axis. Sun elevation angle is approximately 45 degrees.

Description of the area

The Gulf of Guacanayabo is nearly 100 km wide at its entrance, which coincides with the 20 m contour line and the steep slope of the shelf. There is a tidal range of 0.5 m. Tidal streams drive counter-clockwise through the gulf, from east to west at a velocity of approximately 5 nautical miles per day.

Interpretation

There is extreme penetration of light and transparency through the clear water of the gulf, reflecting the bottom to an estimated depth of 10 m. The water of the Gulf is of blue colour. Only in the nearshore areas of the inner gulf there is a screen of grey-pink colour, probably indicating nearshore turbid waters over flat submarine topography.
Depending on the sun's azimuth angle with the exposure axis of the photograph in coincidence, there is no sun glitter on the sea. Of exceptional interest is the area of Gran Banco de Buena Esperanza in the centre of the Gulf, there is a scattered 50 km long network of coral reefs, some covered and others above water. It is not possible to interpret the photograph to tell which reefs break surface. But using this satellite photograph as a whole, it offers for first time the total overview of such an oceanographic-geological phenomenon of a large area. As a result, this reef network can be mapped easily on a small scale map. (Scale of photograph is 1 : 1 million.) Such a map would contain a better delineation of the reef pattern than other maps on the same scale. The interpretation was done in comparison with maps, nautical charts and pilots; interpretation through comparative analysis. Although quantitative bathymetric mapping is not possible, colours and tonal variations could be used to delineate different submarine features.
Concerning remote sensing, this seems to present a simple example, since mapping reefs means mapping geomorphological forms, i. e. it is a static model; there are not many problems other than those inherent in two media photography. Distortion is unimportant here for thematic interpretation (i. e. geometrical-photogrammetric questions).
Transparency and penetration through clear ocean waters is a process of some complexity and the interpreter's difficulties arise (compare brief discussion in the general text and the figure on the topic of water transparency).

Bibliography

Jonin, A. S. et al., 1975, Insular shelves of the tropical ocean: Academy of Sciences of the USSR, P. P. Shirshov Institute of Oceanology, Moskow, 215 p.

Kühlmann, D. H. H., 1970, Die Korallenriffe Kubas, I. Genese und Evolution: Int. Rev. Gesamt. Hydrobiol., vol. 55, no. 5.

Chart ONC – WAC code Y – 26

International map 1 : 1 000 000 N. F. 18 "Santiago de Cuba" and N. F. 17 "Habana"; Issue 1940, revised VIII 1947. American Geographical Society of New York.

Nautical Chart, Deutsche Seekarte, DHI, no. 442, scale 1 : 2 500 000 Ausg. 4, 1971.

Pilots: Westindien Handbuch, III. Teil, DHI.

Water-body of a lagoon consisting of turbid water masses having no transparency

Lagôa dos Patos, Brazil, coast of Atlantic Ocean
Orbital remote sensing from Apollo
and Skylab missions – comparison and evaluation

PÔRTO ALEGRE
Río Guaíba
Douradinho
Ponta das Disertas
BRASILIEN
Tapes
Saco de Tapes
Lagoa dos Patos
pleistozäne Terrasse
Ponta S. Simoa
Arambaré
Río Grande do Sul
Río Camaquã
Tavares
Estrêla
S. Lourenco do Sul
Ponta do Boruju
Bojuru
Pelotas
Wolken
Wolken
Estreito
ATLANTISCHER OZEAN
RIO GRANDE
Querênica
Lagoa Mirim
Taim
Barra do Rio Grande
BRASILIEN
ARGENTINIEN
URUGUAY
Bildausschnitt
1
2
3
4
5
6
7
8
9
10

Lagôa dos Patos, Brazil
Coast of Atlantic Ocean

Interpretation map, from NASA satellite colour photograph

APOLLO mission 7, no. 4–1594

Interpreter H. G. Gierloff-Emden, 1975

Legend of topographic and hydrographic features classified in zones with selected discriminated objects (generalized: no differentiation of distinct water levels on this small-scale map)

Scale approximately 1 : 1 200 000 (varies, caused by oblique photograph)

I above High Water level: terrestrial

mainland on the map as white area

1 margin of dunes and sandy beaches

2 beach ridges

3 inner lagoonal shore with beach ridges and spits

II horizontal extension of inter-tidal zone is lacking or small

III a below Low Water level: inner lagoonal

4 turbid water mass of northern part of Lagoa dos Patos (upper lagoon), yellowish-pink and yellowish-orange (brown) coloured on the satellite photograph, turbid waters, discharged from Rio Guaiba

5 turbid water mass, middle (central) and southern part of Lagôa dos Patos, in motion, with plumes, tongues and eddies,
also area near town Rio Grande and Lagoa Mirim, grayish-yellow and pinkish-grey coloured on the satellite photograph

6 water masses of southern part of Lagôa dos Patos (lower lagoon), more grayish-blue coloured on the satellite photograph; dark patches caused by cloud shadows

III b below Low Water level: marine

7 turbid water mass in front of the Canal do Rio Grande (over Barra do Rio Grande)

8 nearshore and offshore turbid waters with marginal eddies (band of 20 to 30 km wide, over bottom 0 to 20 m deep) grayish-blue coloured; dark patches caused by cloud shadows

9 water of Atlantic Ocean, more light grayish-blue and blue coloured on the satellite photograph, over deeper off-shore area (shelf)

10 boundaries of water masses in motion

Satellite photograph:

Lagôa dos Patos, Brazil,
Coast of Atlantic Ocean

Flight information data:

Mission: *NASA – APOLLO 7*

Date: *12. Oct. 1968*

Photograph: *colour*

Magazine: *4 frame no. 1594, (also 68–HC–694)*

Camera: *Hasselblad 500 C*
70 mm photography

Lens: *Zeiss Planar f/2,8; 80 mm*

Film: *Kodak Ektachrome*

Filter: *none*

Altitude above mean sea level: *195 naut. miles*

Geographical position of the principal point:
latitude: *31 degrees 08 min. S*
longitude: *51 degrees 02 min. W*

Chart: *ONC – WAC code Q – 28 and R – 24*

Local observation time: *17.45*

Sun azimuth: *265°*

Sun elevation: *15° or less*

Photograph viewing mode: *oblique*
exposure axis to 200°

Cloud cover: *13 % alto cumulus, cumulu-stratus*

Scale of the printed picture: *1 : 1 200 000 (approximate)*

Colours of the printed picture (Munsell scale):
The colours of the printed photograph occur more in grayish-pink in comparison to the NASA copy, which shows up more in blue tone.

NASA – APOLLO 7

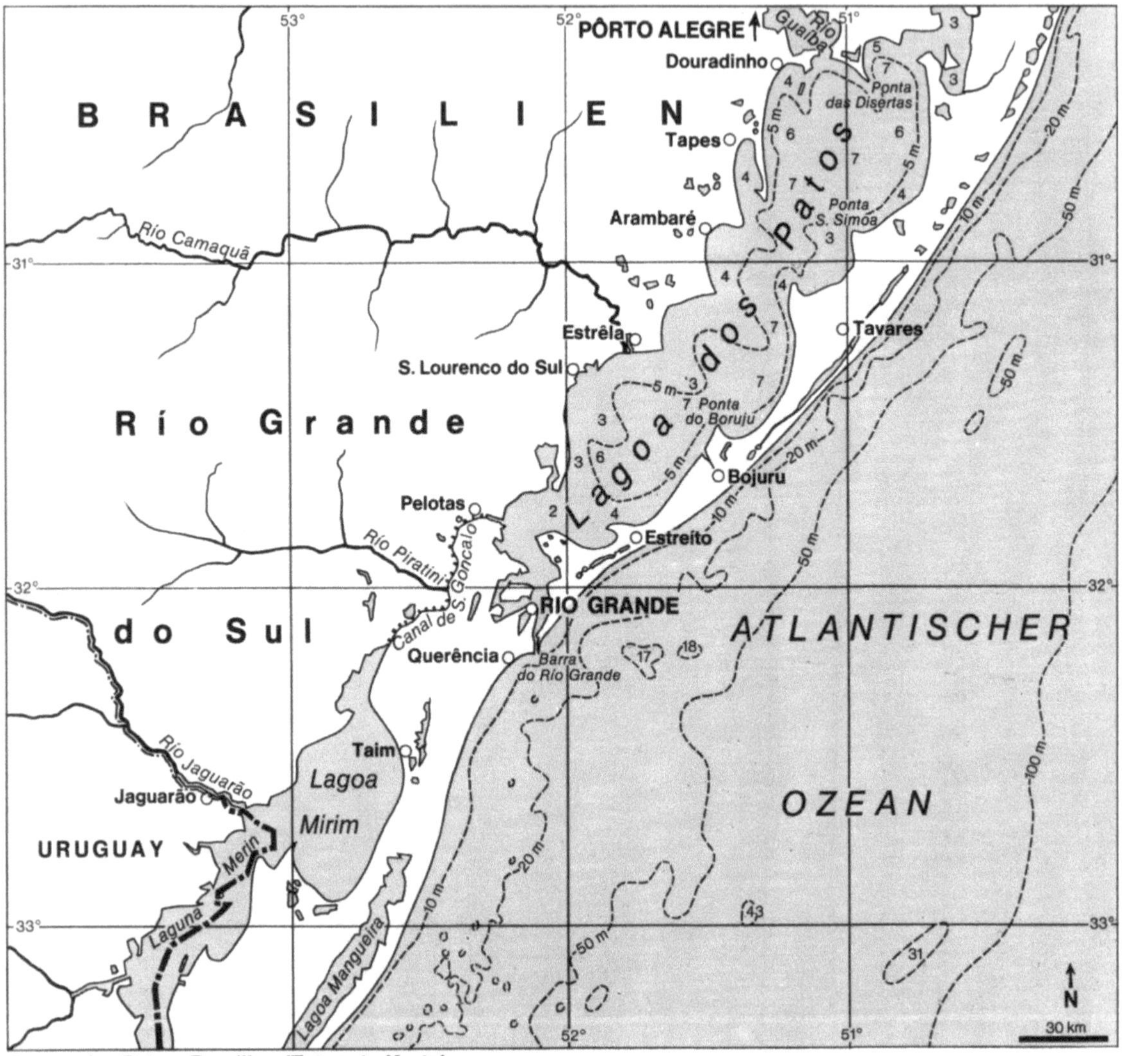

Lagoa dos Patos, Brasilien (Entzerrte Karte)

Lagôa dos Patos, Brazil, coast of the Atlantic Ocean

Chart of the area of the satellite photograph
scale 1 : 2 400 000,
with bathymetric topography, depth in m

Water-body of a lagoon consisting of turbid water masses having no transparency

Lagôa dos Patos, Brazil, coast of Atlantic Ocean
Orbital remote sensing from Apollo and Skylab missions – comparison and evaluation

NASA colour photograph; Apollo mission 7, date 12. Oct. 1968
The scale of this photograph is approximately 1 : 1 200 000 (printed picture scale). The photograph is a slightly oblique one; there is a geographic distortion (southwestern part of the picture is at a relatively smaller scale in comparison with the northeastern part). It was taken at late afternoon time about 18.00 local time, so the azimuth of the sun is westerly. Exposure axis of the camera to the southwest. The sun elevation angle is low, less than 15 degrees. So there is some reflexion in the southwestern sector, i. e. lower left of the picture. Due to the sun's low angle of elevation, there is greater reflexion from, and less penetration through the water.

Lagôa dos Patos of Brazil is one of the largest lagoons on the coasts of the world. The lagoon is 216 km long and 60 km wide. An oblique satellite photograph shows the lagoon. The nature of the landscape and the geomorphological processes are described by Patrick J. V. Delaney, who carried out fieldwork in that area and published two books (see bibliography). The description of the area is reported close to Delaney's papers.

Description of the area

The coastal plain of Rio Grande do Sul, Brazil and north Uruguay is essentially a low sandy area which is characterized by broad flat lowlands, numerous lakes, coastwise terraces, and aeolian land forms. In Rio Grande do Sul one of the world's longest uninterrupted sand beaches extends from the coastal town of Torres south to Uruguay. This beach is at least 640 km long and has a maximum width of 1 km.

The Rio Grande do Sul coastal plain has as its eastern limit the Atlantic Ocean, and as its western limit, the low flat land is bordered by an arcuate belt of higher land which is composed of rocks ranging in age from Pre-Cambrian to Jurassic.

The Lagôa dos Patos is developed, along the eastern margin of the continent where several down-faulted post-Cretaceous coastal basins were formed. One of these, the Pelotas Basin, underlies the coastal plain of Rio Grande do Sul and northern Uruguay and is responsible for the broad, low coastal area.

Geomorphology

"Geomorphologically, the southern Brazil-Uruguay coastal plain is subdivided into five principal physiographic units: a coastwise maritime sand plain including extensive dune fields; a higher, coastwise, Pleistocene terrace plain; an elongate area of low rolling hills and valleys inland from the large lagoons; narrow, paired terraces bordering both sides of the Guaiba estuary near Porto Alegre; and an area of stabilized, dune-like, rolling, sandy hills inland from the terrace plain." Delaney, 1966.

The coastal belt is composed of three roughly parallel strips – flat sand coastal beaches, dune fields, and the lakes along the base of the pleistocene escarpment (4 m to 6 m in height). These lakes are discernable on the satellite photograph. Topographically the Rio Grande do Sul coastal plain consists of a flat lowland whose elevations are seldom more than 6 m.

An important sedimentological problem of the coastal plain is an understanding of the origin of this enormous fine sand mass. The sand originates from the Botucatú sandstone, which is known to outcrop in the northeastern portion of the coast and perhaps crops cut subaqueously on the Rio Grande do Sul swell.

Wind, sediments and topography of the lagoon

The intensity and direction of the wind is considered to be the most important physical process operating in the coastal plain. Aeolian features are aligned parallel to the path of the prevailing northeast wind. Besides this prevailing wind, there are two other accessory winds which have been given names by the local residents – the "Minuano" and the "Carpinteiro da Costa".

"The "Minuano" is a cold, dry, continental, west wind. The "Carpinteiro da Costa" is a southeast or south-southeast wind. It is an on-shore, oceanic wind, and normally blows for three or four days at a time. Many navigation accidents occur when this wind blows because quantities of bottom sand are shifted further out on the continental shelf, changing configuration of shoals or banks, and the force of the wind from the southeast causes the ships to drift in shore." Delaney, 1962.

Morphologically, wind action forms large dune fields along the entire coast, deposits elongate tongues of sand, spreads a thin veneer of sand over the older rocks, modifies the shape and depth of water bodies and hinders the growth of beach vegetation.
There are many well developed dune fields along the beach, they vary between 0.5 km and 6 km in width. These areas are clearly marked in a white-yellow colour on the satellite photograph.

"The two wind-water relationships that are most important are dominant wind direction and fetch. Winds blowing in a constant direction for some period of time over a water body tend to generate waves. These waves have a wider spectrum (relationship between amplitude and period) the farther they are from the initial point where the wind strikes the water body (beginning of the fetch). Thus, where the longest fetch and the greatest intensity of wind directions coincide, the lagoa is deeper and wider. Therefore, since faulting initiated the lake complex and water accumulated, the wind direction and intensity has played the most significant role in determining the configuration and depths within the Lagoa dos Patos (Zenkovitch). Several elongate shoals extend in an east-southeasterly direction nearly halfway across the lagoa from the western shore. These seem to be extensions of continuations of the subdued divides that separate the coastal plain west of the lagoa into a series of broad flat valleys." Delaney, 1962.

Segmentation. Segmentation is a process whereby a lagoon is converted into several smaller parts by spit growth at intervals along the lagoon margins. The efficacy of this process depends on initial lagoon geometry, the availability of beach-building sediments, and the organization and magnitude of energy conditions within the lagoon. Lagôa dos Patos favours segmentation because its long axis runs subparallel with the prevailing northeasterly and southwesterly winds. These winds periodically generate high-energy nearshore circulations that move appreciable sediment loads up and along the shore, as A. R. Orme reported from lagoons on the coast of East-Africa.

The ocean beaches

The beaches are discernable on the satellite photograph. Ocean beaches have been formed along 640 kilometres of coastline in Rio Grande do Sul. These beaches vary in width from 100 m to 1 km (strand line to dune field). The beaches are composed of fine- to medium-grained, well rounded, white to yellowish sand.
The normal beach slope and offshore profile is exceedingly flat. (The beach slope is 2° or 3,4 m in 100 m). Offshore profiles show that the subaqueous slope is 3°27' or 6 m per 100 m.
Sand transport along the coast is initiated due to wave action. The uniformity of the beach is interrupted by breaches or breaks ("sangradouros" in Portuguese) which form during the rainy seasons. These small channels originate during periods of heavy rainfall when the water table within the dune field becomes higher than sea level and eventually breaks through the beach. Water seeps through the beach until it finally begins to scour the sand and transport it to the ocean, causing channels up to a metre deep and several tens of metres wide. These channels exhibit a distinctive braided pattern mainly during the rainy winter months, so there is a water mass transport from the lagoon to the ocean. These features are not discernable on the satellite photograph.

Foreshore and offshore

The foreshore and offshore are located on the so called Rio Grande shelf which is 100 km wide. Close to the Rio Grande inlet there are large sand waves on the bottom of this shelf between the 10 to 50 m contour lines. This submarine topography is not discernible on the satellite photographs.

Marine currents and tides

Two major ocean currents, Falkland (current to north, prevailing June to Aug.) and Brazil (current to south, prevailing Dec. to Febr.), alternately affect the coast of Rio Grande do Sul. Neither of these currents are especially strong.

Other physical factors which are important but almost unknown entities are the storm regime and tidal variation. There are very few storms in Rio Grande do Sul.
The only tide gauge in the area is located about 6 km inland from the jetties at Rio Grande. Here the tide gauge measures the lunar tide plus the apparent wind tide (the effect of the wind on the water). Tide gauge readings over five years indicate that the average tidal range is 1,20 m.
Those tidal streams are responsible for the interchange of lagoon water with ocean water in the area of Rio Grande, flowing through Barra do Rio Grande on the southern end of the Lagôa dos Patos.
In this area, it seems from the photograph, that water from the lagoon, laden with suspended sediment material, flows out to the ocean, where it looks like yellow-grey plumes, to a distance of about 30 km.

Water bodies of the lagoon and water masses; detection from satellite photographs

The water bodies in this coastal area together occupy an area of 13 680 km square, a size comparable with Lake Maracaibo in Venezuela, (Lagôa dos Patos 9 910 km square, Rio Guaiba 500 km square, Lagôa Mirim 3 770 km square). Generally, water depth is proportional to lake size. Near the middle the Lagôa dos Patos is only 8,7 m deep.

"The Lagôa dos Patos consists of three distinct parts: The lower Lagôa dos Patos between the Barra do Rio Grande and the Barra da Feitoria, the central Lagôa dos Patos from Barra da Feitoria to a point near Itapõa, and the upper Lagôa dos Patos consisting of the northeastern part and the pass north of Itapõa into the Guaiba estuary proper." Delaney, 1962.

The lower lagoon is the most difficult to navigate because of its depth variability and currents. Considerable quantities of flocculated fine sediments are constantly being deposited, causing rapid channel fill which necessitates the continued use of dredges.
This part cannot be differentiated on the photograph, but several plumes of water can be detected, consisting of "lower lagoon water", remixed with ocean water. There are some dark patches in this area, shadows caused by clouds.

The central lagoon. "The central Lagôa dos Patos is the most stable of the three subdivisions of the lagôa, and consequently is commonly navigated by ships drawing up to 20 feet of water. The bottom of this part of the lagôa is floored principally with arkose, and does not require dredging." Delaney, 1962.

This part is well differentiated on the satellite photograph. Although because of some transparency effect, water depth may have an influence here, it seems sure that the sediment laden water from the Rio Camaqua enters the lagoon in form of yellow-brown coloured eddies covering the southern half of the central Lagôa dos Patos.
The northern half of the central Lagôa dos Patos shows on the photograph in a light brown colour, lacking any differentiation, but is different from the southern area as mentioned. These water masses are probably pure fresh water, as is the northermost water body.
The upper lagoon. The northern part of the lagoon is filled with very turbid water masses. The Guaiba estuary which joins the Lagôa dos Patos in the northwest, does partly occur within the frame of the photograph. Discharge rates of the rivers with sediment laden waters are important but virtually unknown. Concerning this area and phenomena, compare with the Skylab photograph.

Water circulation in the Lagôa dos Patos

"The Lagôa dos Patos has a very low gradient averaging about 1 m in 120 km. Because of low gradient and relatively minor volume of run-off, waters of the lagôa are almost at a standstill, and wind velocity and direction play a dominant role in the internal dynamics of the lagoons. Distribution of salinity, for example, is primarily controlled by wind direction and intensity. In the Lagôa dos Patos, when strong winds blow from the southeast, brackish water (salts up to 2 ‰) may extend as far north as Itapõa. However, when the wind blows from north or northeast, fresh water may be encountered between the jetties at Rio Grande. Thus, the Lagôa dos Patos is brackish when the winds are from the southeast and fresh when the winds are from the northeast." Delaney, 1962.

This accounts for the flat bottom-gradient and low water-exchange rate of the entire lagoonal complex. This system is considered lacustrian because almost all precipitation within the basin passes through an intermediate lake or lagoon before entering the Atlantic Ocean. The lagoons, Lagôa dos Patos and Lagôa Mirim, are large shallow bodies of water that are connected to the South Atlantic Ocean by a single inlet at Rio Grande. The inlet at Rio Grande has a sufficiently deep channel (12 to 16 m), to permit large ocean vessels to enter the port of Rio Grande. The channel is maintained artificially as sediment deposition is continuous. Within the inlet salinities vary radically (between 1 and 34 ‰), and when sediment laden fresh waters from the Lagoa dos Patos come in contact with saline to brackish ocean waters, suspended clays flocculate and settle to the bottom.

Conclusion: Remote sensing with satellite photography

The Lagôa dos Patos is a very large area, where continuous fieldchecks and aerial photography are costly and practically neither possible nor economic.
Physical agents of significance within this area are: climate, dynamics of interior water bodies, nearshore hydrography, currents, waves and tides, and river discharge.
Some phenomena can be detected on the satellite photograph; if these are stages of processes, as plumes of water masses with suspended sediments in form of plumes or eddies, the satellite photograph can give additional information for thematic mapping.
For further information there is a need for a continuous series of satellite photographs with an adequate repetition rate in relation to these processes.

Bibliography

Barcellos, B. N. and J. D. Richardson, 1960, Carta pesqueira do Rio Grande do Sul, pesquisa sôbre a pesca do alto mar em Rio Grande: *Secr. Agric.*, Porto Alegre.

Delaney, P. J. V., 1962, Quaternary geologic history of the coastal plain of Rio Grande do Sul: *South American Coastal Studies*, Contr. no. 62–6 (December), part A Coastal Studies Inst., Louisiana State Univ., Baton Rouge.
The second part of the eighteenth in a series of reports prepared under project no. Nonr 1575(03), task order no. NR 388 002 of the geography branch, Office of Naval Research.

Delaney, P. J. V., 1966, Geology and geomorphology of the coastal plain of Rio Grande do Sul, Brazil and Northern Uruguay: *South American Coastal Studies*, Tech. Rept. no. 18, Contr. no. 66–6, part B.

Ihering, H. V., 1885, Die Lagôa dos Patos: *Deutsche geogr. Blätter*, 8 (2), p. 163–204.

Phleger, F. B., 1967, Some general features of coastal lagoons: in Lagunas Costeras, Symposium UNAM, UNESCO, p. 5–26.

Urien, C. M. and M. Ewing, 1974, Recent sediments and environment of southern Brazil, Uruguay, Buenos Aires, and Rio Negro continental shelf: in C. A. Burk (ed.), The geology of continental margins, p. 157–177.

Vogt, H., A. Gomes and J. Tricart, 1960, Note preliminaire sur la morphologie du cordon littoral actuel entre Tramandaí et Tôrres, Rio Grande do Sul, Brésil: *Cahiers Oceoanogr.* du C. O. E. C., vol. 12, no. 7, p. 453–457.

Zenkovitch, V. P., 1959, On the genesis of cuspate spits along lagoon shores: *Jour. Geol.*, vol. 67, no. 3 (May), p. 265–277.

Maps of Lagôa dos Patos

Several good general nautical charts of the coast of Southern Brazil and Uruguay have been made by the Marinha do Brasil, Diretoria de Hidrografía e Navegação, and the Uruguayan Navy. The most useful charts are Brasil Cartas 1, 30, 90, and 2101. The Departamento de Portos, Rios e Canais do Ministerio de Viação e Obras Públicas have published several good maps: Lagôa dos Patos, Pôrto de Rio Grande, Rio Guaíba, Delta do Rio Jacuí, and the Ship Canal between Rio Grande and Pôrto Alegre.
In 1948 and 1951, world aeronautical charts numbers 1384 and 1433 were released. Later, in 1954, a revised edition of chart 1384 in Portuguese was published.
Three geological maps of the coastal plain of Rio Grande do Sul are included in an unpublished doctoral dissertation in the geology library of the Universidade de São Paulo (Delaney, 1962). The maps include the entire coastal plain of Rio Grande do Sul and northern Uruguay (two sheets, scale 1 : 500 000), a map of the Osório homocline (1 : 100 000), and a map of the Rio Grande area (1 : 100 000).

Charts ONC – WAC code Q – 28 and R – 24

Nautical charts

Deutsche Seekarte DHI Nr. 945, in scale 1 : 300 000, issue 1967 "Tramandaí bis Rio Grande mit Lagôa dos Patos", with 5 m depth contour in the lagoon.

Weltkarte 1 : 2 500 000, DDR, 1960, SG-I-22-24, no. 170.

Carto do Brasil; "Porto Alegre", Folha SH 22, and "Lagôa Mirim", SI-22, 1 : 1 000 000, issue 1959. (has many place names, no depths in the lagoon).

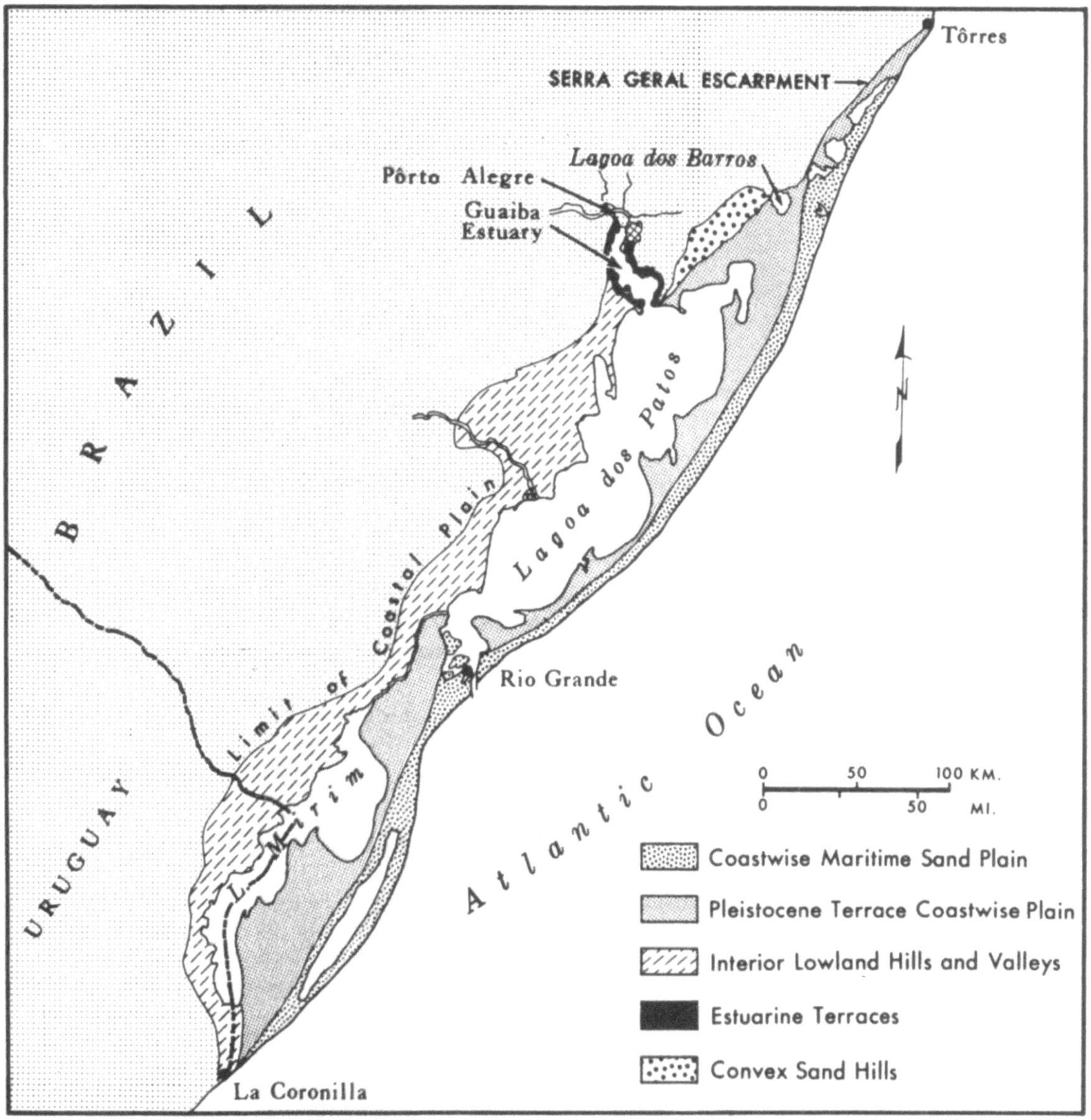

Physiographic units in the coastal plain of Rio Grande do Sul.

Skylab multispectral camera photograph, no. SL 3-28-331, Febr. 1974
(published in Skylab Earth Resources Data Catalogue, NASA 1974, p. 109),
for comparison with Apollo 7 photograph; an example of evaluation from orbital remote sensing products of NASA.

This Skylab photograph is a vertical one and presents more details due to the better ground resolution, but it presents less details in coastal water features.

In this publication the picture is presented as a black and white copy of the Skylab colour photograph.

Lagôa dos Patos in southern Brazil

River discharge of sediment laden water into Lagôa dos Patos from Rio Guaiba, marked on the photograph as "SL 1".

This photograph presents in detail the circulation pattern, the dumping of sediment and depositional features Specific areas of beach erosion and deposition can be identified on the cuspate spits in the lagoon (compare text). This heavy sediment laden river discharge is significant mainly for the upper lagoonal water mass. Dynamic of river discharged waters form an eddy of 25 km diameter with clockwise vortex. It is remarkable that the sediment laden waters of the river form a closed band on the north side of the river and do not mix in the Rio Guaiba estuary with waters from the southern banks of the river. This is an effect which could be observed by remote sensing technique in other rivers, as for the rivers Saar and Rhine in Germany (S. Schneider, 1974, Gewässerüberwachung durch Fernerkundung – Die mittlere Saar: Landeskdl. Luftbildausw. im mitteleurop. Raum, Schriftenfolge Bundesforschungsanst. Landeskde. Raumordng., vol. 12, Bad Godesberg.).

Athmospheric — biological phenomena and the bottom as a boundary layer of a very flat lagoon

Laguna Madre de Tamaulipas, and Gulf of Mexico

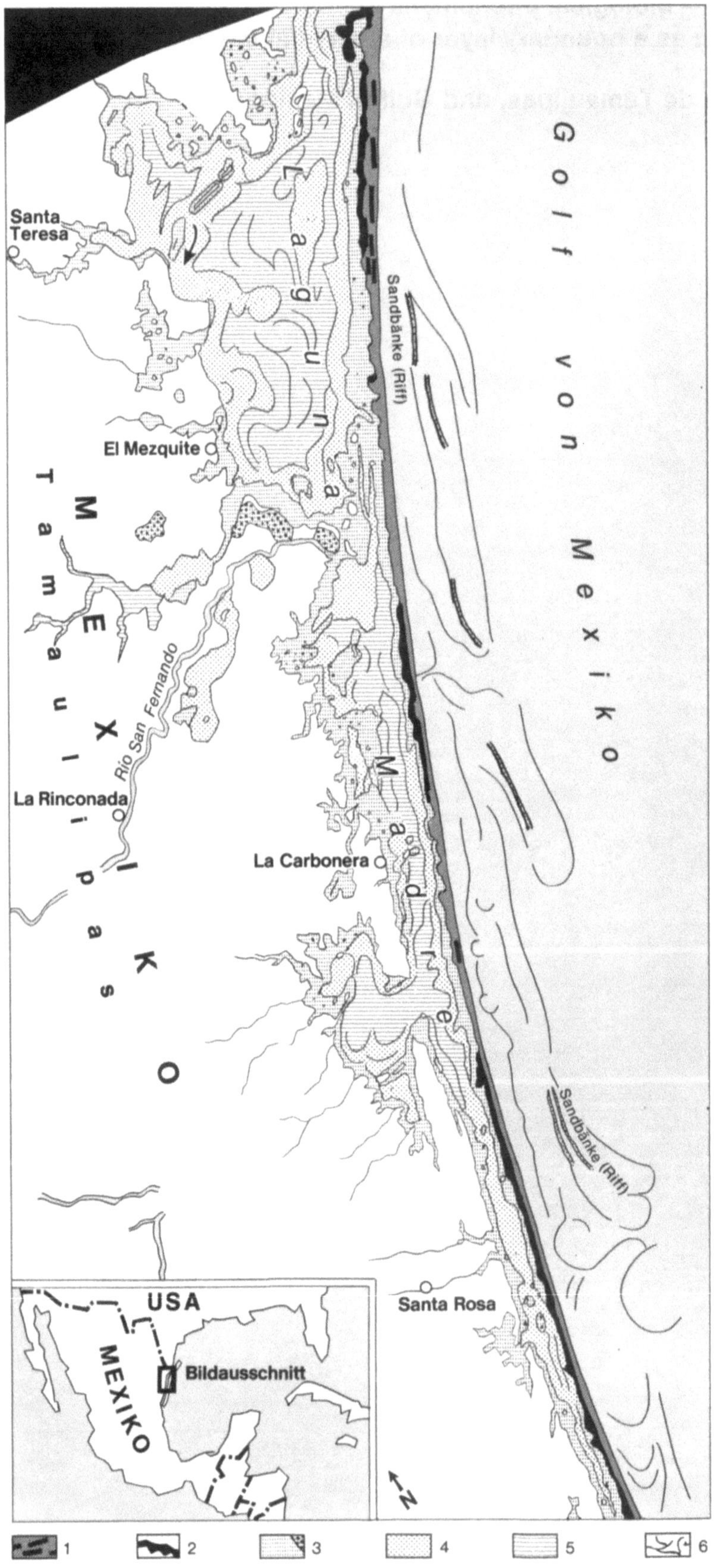
Golf von Mexiko
Laguna Madre
MEXIKO
Tamaulipas
Santa Teresa
El Mezquite
Rio San Fernando
La Rinconada
La Carbonera
Santa Rosa
Sandbänke (Riff)
Sandbänke (Riff)
USA
MEXIKO
Bildausschnitt
1
2
3
4
5
6

Laguna Madre de Tamaulipas, Mexico and Gulf of Mexico

Interpretation map, from NASA satellite colour photograph

APOLLO mission 7, no. 5–1633

Interpreter H. G. Gierloff-Emden, 1975

Legend of topographic and hydrographic features classified in zones with selected discriminated objects (generalized: no differentiation of distinct water levels on this small scale map)
Scale approximately 1 : 800 000

I above High Water level: terrestrial

mainland on the map as white area

1 barrier island with dunes and beach (sand) with some beach ridges

2 sand ridges and wash-over fans (sand and loam), partly with vegetation, on the lagoonal side of the barrier island (1 and 2: coastal barrier complex)

II between High and Low Water level: inter-tidal zone

3 flats, temporarily exposed at low water (caused by tidal streams and wind driven currents) partly wind-tidal flats

III below Low Water level: lagoonal and marine

4 flat areas of lagoonal bottom covered with water (0,3 – 1 m deep), mostly turbid water

5 turbid water masses, partly in motion

6 boundaries of water masses in motion (stream-lines, tongues and eddies)

Satellite photograph:

Laguna Madre de Tamaulipas, Mexico and Gulf of Mexico

Flight information data:

Mission: *NASA – APOLLO 7*

Date: *12. Oct. 1968*

Photograph: *colour*

Magazine: *5* . *frame no. 1633*

Camera: *Hasselblad 500 C*
70 mm photography

Lens: *Zeiss Planar f/2,8; 80 mm*

Film: *Kodak Aerial Ektachrome*

Filter: *Wratten 2 a*

Altitude above mean sea level: *123 naut. miles*

Geographical position of the principal point:
latitude: *24 degrees 45 min. N*
longitude: *97 degrees 30 min. W*

Chart: *ONC – WAC code H – 23 and H – 24, J – 24 and J – 25*

Local observation time: *12.00*

Sun azimuth: *180°*

Sun elevation: *60°*

Photograph viewing mode: *vertical*

Cloud cover: *30 % cumulus, altocumulus*

Scale of the printed picture: *1 : 800 000*
(section of satellite photograph)

Colours of the printed picture (Munsell scale):
The colours of the printed photograph occur more yellowish and greenish yellow in comparison to the colours of the NASA copy, which shows up less yellowish, but more bluish and purplish pink.

NASA – APOLLO 7

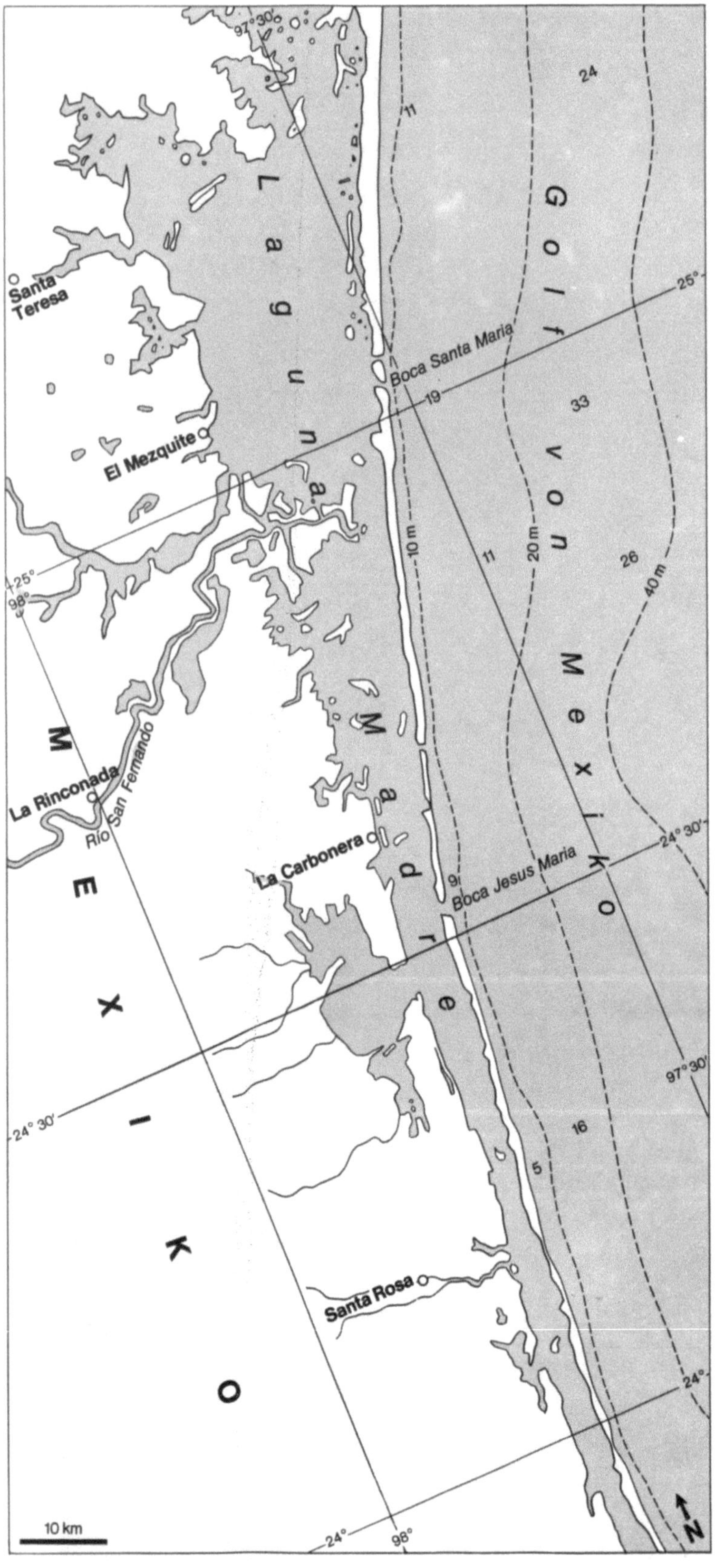

Laguna Madre de Tamaulipas, Mexico, Gulf of Mexico

Map of the area of the satellite photograph scale 1 : 770 000 (approximate), with bathymetric topography

Athmospheric – biological phenomena and the bottom as a boundary layer of a very flat lagoon

Laguna Madre de Tamaulipas, and Gulf of Mexico

NASA colour photograph, Apollo mission 7, AS 7–5–1633; date 12. Oct. 1968, Mexico, Gulf Coast; hand operated camera. The scale of this photograph is approximately 1 : 800 000 (printed picture scale). The photograph is an approximately vertical one, geographic distortion is small. It was taken at noontime, about 12.00 local time, so that the azimuth of the sun is southerly. The sun's elevation is approximately 60 degrees.

The satellite photograph of the Laguna Madre area of the Mexican coast on the Gulf of Mexico is certainly a wonderful picture of a landscape of our earth, showing a lagoon as a margin of land to the sea.
No map exists on a scale close to that of this photograph, neither do the international (1 : 1 million) nor nautical charts present such a detailed picture of this large area.
But the question is, what does this coloured pattern represent by way of geomorphological and oceanographic objects or as a stage of hydrological processes, at that very moment when the film of the camera was exposed?
The intent of this paper is, to analyse the information that can be extracted from this photo and to show what additional information is available compared with maps prepared before this photo was taken, as well as what can be gained by using satellite photography as a tool and method for mapping.
As a description of this area of the Laguna Madre shows, this very shallow and extended lagoon includes so many different phenomena and processes, that it is not generally possible to identify colouration in the picture as a particular object.
In this very shallow lagoon, plant life plays an important role in colouration of both the bottom and the water masses.
"Few areas offer such a fascinating picture of continuously changing conditions of sedimentation or provide such a clear record of changes which have occurred in the recent geologic past." Fisk, 1959.

Description of the area

Along the west coast of the Gulf of California, north and south of the Rio Grande delta, a practically continuous barrier island chain, backed by a lagoonal depression, separates the mainland from the Gulf of Mexico and gives a distinctive character to the region. South of the Rio Grande, an area of the Mexican state Tamaulipas, the flat Laguna Madre extends for a distance of about 250 km. The Laguna Madre de Tamaulipas extends from about 25°30'N to 23°30'N and has an area of about 1 000 square kilometres. Hildebrandt, 1958. Mean water depth is only 1 m.

The barrier island

The barrier island, Madre Island, is narrow, only some 100 m to 2 km wide. On the barrier island, a complex of small dunes covers extended areas; they are mostly not more than 5 m high.
Sands drifting inland from the berm form a narrow belt of active foredunes with an irregular western front which blankets the fixed dunes of the island ridge.
Although the individual dunes are below the scale of minimum visibility on the satellite photograph, the geomorphological formation as a whole is well recognizable. In the northern part of the barrier island, the dune fields are broader, in the southern part though they are very small. In some places sand bulges occur on the lagoonal side as extensions of the dune field (see photograph). In the southern part, a very small brown band along white coastal dunes probably represents clay dunes. Ephemeral active dune fields in the washover apron change the feature of the island's western margin.
Madre Island exhibits a remarkably smooth and unbroken beach and with its combined length of 210 km, it is one of the longest barrier islands in the world. The Madre Island beach is composed of sand and shell and is

about 100 m wide. On the photo it occurs as a white line, well differentiated against the dunes.
The barrier island separates the lagoon from the ocean. Due to the divergence of the barrier island from the mainland, the northern part of the Laguna Madre (close to Rio Grande delta) is wider than its southern part.
On maps the barrier island is cut through by a number of channels. The satellite photograph does not show such features, but such gaps certainly exist at certain times of the year.
Several gaps in this section of the ridge have been scoured to below-sea-level depths. Gaps are periodically flushed out by the action of storm-driven waters, but, where their entrances have subsequently been sealed by beach deposits, many of them hold ponds at the back of the beach.

The water bodies of the lagoon

The Laguna Madre depression holds a series of elongate shallow lagoonal basins joined by above water flats of sand and mud. The lagoon is flat, in extended areas less than 1 m deep and falls dry in parts during the summer season of the year, especially in its southern part. The northern part begins where the river Rio San Fernando whose flats rise 1 m above sea level, has built up a small delta in the lagoon. The northern part has a distinct water body. Slight but significant differences in elevation determine the frequency with which an area is flooded.
The eastern section of the flats merges with the hummocky dune terrain of Madre Island.
The western section occupies an erosional re-entrant of the mainland margin, which extends inland for approximately 2 to 15 km and incorporates island-like remnants of the mainland plain.

Laguna Madre and delta of the Rio Grande

The Laguna Madre in its geologic history depends on the development of the Rio Grande delta. Three distinct stages of sub-deltas have been identified by geologists. The northern part of the present lagoon borders one of the former sub-deltas.
Sediments contributed by the former courses of the river gradually filled the northern end of Laguna Madre.
To the north of the lagoon, some abandoned channels of the Rio Grande are visible. The Spanish term "resaca" is applied to abandoned tributaries of this delta. Hildebrand notes the presence of oyster shell deposits there which he suggests may have been accumulated at a time when the Rio Grande carried more water and entered the Gulf of Mexico in Tamaulipas instead of Texas.
Although it is probable that sediments contributed to the Gulf by the Rio Grande have been and are currently being swept both in a northerly and southerly direction along the coast, a substantial portion of the material has been deposited in the immediate vicinity of the deltaic plain of the Rio Grande river.
The local filling of the lagoonal depression and the growth of Madre Island have proceeded contemporaneously in late recent time through the interaction of aeolian, aqueous, and biologic processes.
During the period from 1929 to 1943, the Rio Grande transported more than 10 million tons of silt down to the delta.
Because large amounts of river water have been diverted to irrigate the fertile lowlands along its course, the mouth of the river ordinarily has a small flow. Occasionally the lower Rio Grande dries up completely (because of fertile soils and mild climate, the delta area or "Lower Valley", as it is locally called, has been changed by irrigation from a desert to a rich agricultural area). As a result, the filling of the Laguna Madre has changed due to these irrigation programmes. Thus, the Rio Grande transports a small load of sediment to the Gulf.

Tide, winds and currents, salinity and sediments

Consequently, the tidal range in Laguna Madre is only a few centimetres and may be diurnal, semidiurnal or mixed. Wind effects are often more important than the tide in this shallow lagoon. The wind tides produce changes in water level, predictions in excess of 1 m are based on observed wind velocities. Slight but significant differences in elevation determine the frequency with which an area is flooded. This topography and the wind cause so-called wind-tidal flats.

Winds also control the currents in Laguna Madre; winds from the north cause currents to set southward, winds from the south cause currents to set northward.
Over large areas the lagoon is so shallow that a change in the wind direction can cause the water to disappear, leaving broad mud flats. During winter some water is moved into the southern lagoon by the action of northerly winds (Northers).
In places, the salinity of the Laguna Madre is sometimes well above that of the ocean water of the Gulf of Mexico, although the gulf coastal waters in this area have as high as 36 ‰ salinity, in comparison with the gulf coastal waters in the Galveston area, where it is around 32 ‰. The lagoonal water sometimes has even higher salinity. Cloudbursts in the area bring about abrupt reductions of surface salinity to as low as 2 ‰.

The high salinity of this lagoon is a product of heavy evaporation, as well as of the scarcity of streams the evaporation exceeds river runoff plus rainfall. During the long dry season (summer), portions of the very shallow lagoon entirely evaporate, and fine clays from the floor are blown away, leaving circular depressions. The basins thus excavated usually fill with water during the rainy season.
The muds of the dry beds of the lagoon contain precipitated salts, such as gypsum. Clay particles are stuck together by the salts, and small aggregates are blown like sand grains and accumulate as clay dunes at the borders of the basins. These clay dunes are probably recognizable as small brown lines along the margin.
Gypsum, that has been precipitated from sub-surface waters, and mat-forming blue-green algae, which have flourished on sections of the flats and influenced sedimentation, reflect changing salinities.
Clays make up a substantial fraction of mud-flat deposits and are carried in suspension by lagoonal waters.

Colour of lagoonal water and bottom, depending on biological processes and time scale

The plant life of the Laguna is highly seasonal, dying down during the high summer temperatures in August and September, and remaining dormant until spring. The macroscopic algae begin to appear in February and the grasses start to grow in March. The development of algae in the lagoon is very important, because the colour of the water, and the periodical dry bottom, change in relationship to this hydrobiological phenomena.

Diatomeae migrate out of the bottom substrate up to the surface during daytime low tide so that algae mats form on the dry fallen wadden area, i.e. this process happens in a special time scale. This is important for orbital remote sensing repetition rate concerning target object.

Under hypersaline conditions, luxuriant growths of blue-green algae appear. In certain areas red algae also occur and gives the water a rust-red colour.
Fisk describes the Laguna Madre of Texas, which is slightly similar to the Laguna Madre of Tamaulipas:

"When the water is extremely saline in certain areas such as the North Cove, a red algae, Laurencia, also abounds and gives the water a rust-red colour. As soon as the surface of the western flats is wetted, the dominant algae form, Microcoleus chthonoplastes, flourishes and soon forms a dense mat over the entire surface. The living mat is green to greenish-brown in colour, and its dense velvetlike surface is composed of minute filamentous organisms which extend a few millimetres into the water; clays settling from the water are intermixed with the algae carpet. When the water retreats, the exposed algae become dormant and the surface takes on an earthy, reddish-brown hue. Fisk, 1959.

The grass distribution on the flat bottom of water bodies of 1 m in depth depends on chlorophyll production, i.e. sunlight time and intensity, which is in that area 10 times greater in summer than in winter.

The littoral

The coast at and near the mouth of the Rio Grande has north-east winds from October to January, east winds in March and September, and south-east in February and from April to August.
Prevailing direction of offshore and littoral surface currents throughout the year is in a northerly or northwesterly direction. The velocity of currents varies, but often is of the scale of 25 nautical miles per day, i.e. 0,6 metres per second. The net effect is a northward drifting of the sediments.
The submarine bottom of the shore slopes gently offshore to 16 m below sea level in approximately 1 km distance from the coast. A series of parallel offshore bars which rise up to a metre above the gulf floor locally interrupts the shore-face slope.

Evaporites

Lagoons, estuaries and supratidal flats are features, where evaporites such as salt and gypsum develop as a result of the concentration of seawater by evaporation, i. e. a process of diagenesis and geochemistry to superhalinity in these features often occurs.
Many salt depositions on our continents of today had once been developed in lagoons at the edge of the sea.
This was discovered in the 18th century, when Bischoff reported that these rock types had been found in a bay behind a bar of a lagoon, over which the sea poured only during periods of high tide.
Recent studies of lagoons and tidal flats such as "sabkhas" in the Persian Gulf, mention this process, for instance the study of F. B. Phleger on the Ojo de Liebre lagoon on the west coast of Baja California.
Because salt is of some importance for commercial exploitation it may be pointed out that some of these gypsum or salt flats are identifiable on satellite photographs and imageries.

Photo interpretation

On the satellite photograph, in the middle, there are some few differentiations in the blue colour of the water of the Gulf of Mexico. Light screens of longitudinal and stripe-form of a brighter, more grey-blue colour do occur, mostly nearly parallel to the coast. These could be interpreted as offshore sand bars in a herring bone pattern. But this pattern has such an orientation, that it could be produced by an undercurrent, which is flowing from north-east to south-west, i. e. with an angular to the surface current. This would represent newly discovered submarine topography and a current due to interpretation of this satellite photograph.
The cloud- and furrow-forms in the southern part of the gulf water, of greenish-blue colour, seem to be circulating turbid water masses with some material in suspension.

Conclusion

Because some areas of the lagoon are completely dry seasonally and exposed to the atmosphere, there is an important influence of atmospheric agences such as sun, evaporation etc., on the bottom surface of these flats, i. e. an interaction on the boundary layer of the atmosphere and the (submarine) bottom. Thus chemical processes such as oxidation and reduction have an influence on the surface of the sediments and discolouration or tanning produce a coloured mosaic of small and medium scale patterns on the landscape.
This phenomenon is somewhat similar to the wadden areas, which are periodically exposed to the air, caused by tidal rhythm. Moreover, these biological processes are going on, as development of blue-green algae growth on the mud flats, caused by the hypersaline water mass condition; this often colours the waters of the lagoon. When exposed to the air, these algae produce crusts of different colours. Under extreme salinity conditions, in certain areas, red algae abound and give the water a rust-red colour.
A number of intriguing life cycles have not been worked out and many others are inadequately known.
With respect to the specific colouration on the photograph, printed in this paper, and depending on copy and printing processes, such a rust-red colour can be recognized in the northern part of the area, not very far from "Santa Teresa". But because field work (ground check) is lacking, this is only a possible explanation.
Without a ground check in this area, which must be made simultaneously with the satellite photograph, it is impossible to detect all geomorphological or visible hydrological features of the lagoon with such a single satellite photograph. Satellite photographs taken with a longer repetition sequence, for instance two each month, as with ERTS-1, or now Landsat 2 satellite, could help to complete our understanding of the processes of this lagoon.

Pollution problem of Laguna Madre

Very close to Laguna Madre are the oil fields of Reynosa – Matamoros – a dangerous situation. Any crude oil pollution which pours into this flat lagoon would have a lethal effect on the ecology, because, on dry fallen areas the oil covers the exposed bottom as a film. Moreover there is very little exchange with Gulf waters.

Bibliography

Remark: A large number of publications exist which deal with the Laguna Madre of Texas, i. e. the area north of the Rio Grande delta. The hydrographic and geomorphological features of the two lagoons are very similar.

Emery, K. O., R. E. Stevenson and J. W. Hedgpeth, 1957, Estuaries and lagoons: in J. W. Hedgpeth (ed.), Treatise on marine ecology and paleoecology: *Geol. Soc. America*, Mem. 67, p. 673–750.

Fisk, H. N., 1959, Padre Island and the Laguna Madre flats coastal South Texas: in 2nd Coastal Geography Conference, Coastal Studies Institute, Louisiana State University. (Contains an extended bibliography).

Gunter, G., 1967, Some relationships of estuaries of the fisheries of the Gulf of Mexico: in *Estuaries*, p. 621–638.

Hedgpeth, J. W., 1967, Ecological aspects of the Laguna Madre, a hypersaline estuary; in *Estuaries*, p. 408–419.

Hildebrandt, H. H., 1958, Estudios biologicos preliminares sobre la Laguna Madre de Tamaulipas: *Ciencia*, Mexico, p. 151–173.

Kirkland, D. W. and R. Evans, 1973, Marine evaporites, origin, diagenesis and geochemestry: *Benchmark Papers in Geol.*: Dowden, Hutchinson and Ross, Stroudsbury (Pennsilvania).

Miller, J. A., 1974, Facies characteristics of Laguna Madre wind-tidal flats: in R. H. Ginsburg (ed.), Tidal deposits, p. 67–73.

Milliman, J. D., 1974, Marine carbonates: Part 1 of Recent sedimentary carbonates: Springer, Berlin-New York.

Palmer, J. D., 1975, Biological clocks of the tidal zone: *Scientific American* (February), p. 70–79.

Phleger, F. B., 1965, Sedimentology of Guerrero Negro Lagoon, Baja California, Mexico. In: Submarine geology and geophysics, *Colston Research Soc.*, 17th symposium, London, Butterworth (Colston Papers), vol. 17, p. 205–237.

Phleger, F. B. and G. C. Ewing, 1962, Sedimentology and oceanography of coastal lagoons in Baja California. *Geol. Soc. Am.* Bull. 73, p. 145–182.

Poggie, J. J. jr., Coastal pioneer plants and habitat in the Tampico region, Mexico: *Tech. Rept. Coastal Studies Inst.*, Louisiana State University, Baton Rouge.

Rusnak, G. A., 1960, Sediments of Laguna Madre, Texas: in Recent sediments, northwest Gulf of Mexico, 1951–1958: *Amer. Assoc. Petroleum Geologists*, p. 153–196.

Thompson, D. E. et al., 1973, Seasonal aspects of remote sensing coastal resources: Proceed. 2nd Ann. Remote Sensing Earth Res. Conf., Tullahoma, Tenn., p. 1201–1249.

Trowbridge, A. C., 1932, Tertiary and quaternary geology of the lower Rio Grande region, Texas: *United States Geological Survey* Bull. 837, 260 p.

Wittich, E., 1912, Die Salzlager am Ojo de Liebre an der Westküste von Nieder-Kalifornien. *Zeit. Mineral.*, p. 25–32.

Maps:

Charts ONC – WAC code H – 23 and H – 24, Y – 24 and J – 25

World Aeronautical Map, no. 522, Madre Lagoon, in scale 1 : 1 000 000.

Nautical chart: Deutsche Seekarte, DHI, no. 529, Golf von Mexico, nordwestlicher Teil, issue Aug. 1969, on scale 1 : 1 000 000. (This chart does not show the exact features of the Laguna Madre).

Remark: Neither map or chart shows depth measurements in the lagoon.

Water mass interaction, lagoonal-oceanic, visible due to sediment laden waters

Change detection of processes on diurnal time scale and comparison of Gemini photograph from 1965 with Skylab photograph from 1965 of this area, demonstrating the long term significance of transient processes

Laguna de Terminos and Campeche Bay, Gulf of Mexico

Gulf clear blue water
Golf von Mexiko
Gulf offshore turbid water
Gulf offshore clear water
Golf von Campeche
Barra
Barra de Puerto Real
Isla Aguada
Puerto del Carmen
Ciudad del Carmen
Isla del Carmen
Laguna del Puerto Rico
Punta Zacatal
Laguna de Terminos
Laguna del Este
Rio Palizada
Laguna de Panlau
Rio Candelaria
MEXIKO
Campeche
USA
MEXIKO
Bildausschnitt
N
1
2
3
4
5
6
7
8
9
10a
11
12
13
10b

Laguna de Terminos and Campeche Bay, Gulf of Mexico

Interpretation map, from NASA satellite colour photograph

GEMINI mission 5, no. MSC-S 65-45765

Interpreter H. G. Gierloff-Emden, 1975

Legend of topographic and hydrographic features classified in zones with selected discriminated objects (generalized: no differentiation of distinct water levels on this small-scale map)
Scale approximately 1 : 600 000

I above High Water level: terrestrial

mainland on the map as white area

1 barrier island with dunes and beach (sand), sand ridges

2 inner lagoonal shore (beach)

3 beach ridges

4 salt marshes and marshes

II between High and Low Water level: inter-tidal zone

(4) (compare I) partly mangrove swamp

5 sand bars (barra), for instance delta in the lagoon south of Canal Puerto Real, "tidal delta" partly temporarily exposed or covered

6 swamps, for instance in Rio Palizada area

(12) (compare III)

III below Low Water level: lagoonal and marine

7 water bodies of estuaries of rivers, which discharge into Laguna de Terminos

8 debris laden water mass of Laguna del Este

9 water masses of inner lagoon, near river estuaries, brackish

10a sediment laden, lagoonal water mass, with sand and silt and material in suspension

10b water mass, pouring out of Laguna de Terminos channels into Campeche Bay during ebb tide stream

11 boundaries of water masses in motion, partly stream lines, tongues and eddies (different water masses, as described in this map; arrows as symbols for general current direction)

12 sand bars and shoals

13 sandy reefs of the foreshore zone, nearshore circulation (surf belt) with longshore current

14 marine water masses as described on the map

5 and 12 appear in categories II and III, both as a geomorphologic phenomenon of both levels.
The coastal water mass between beach and "Gulf offshore clearwater" includes 10a, 10b and 13.
8, 10a and 10b are hydrographic features, but occur on the satellite photograph in the same colour and tone as 12, a topographic feature: light grey and grayish blue.

Satellite photograph:

Laguna de Terminos, Campeche Bay and Gulf of Mexico

Flight information data:

Mission: *NASA GEMINI 5*

Date: *22. Aug. 1965*

Photograph; *colour*

Magazine: *MSC – S 65*   *frame no. 45765*

Camera: *Hasselblad 500 C*
70 mm photography

Lens: *Zeiss Planar f/2,8; 80 mm*

Film: *Ektachrome MS*

Altitude above mean sea level: *100–200 naut. miles (mission)*

Geographical position of the principal point:
latitude: *18 degrees 45 min. N*
longitude: *91 degrees 35 min. W*

Chart: *ONC – WAC code J – 25*

Local observation time: *14.00*

Sun azimuth: *270°*

Sun elevation: *65°*

Photograph viewing mode: *slightly oblique*

Cloud cover: *10 % alto cumulus*

Scale of the printed picture: *1 : 600 000*

NASA GEMINI 5

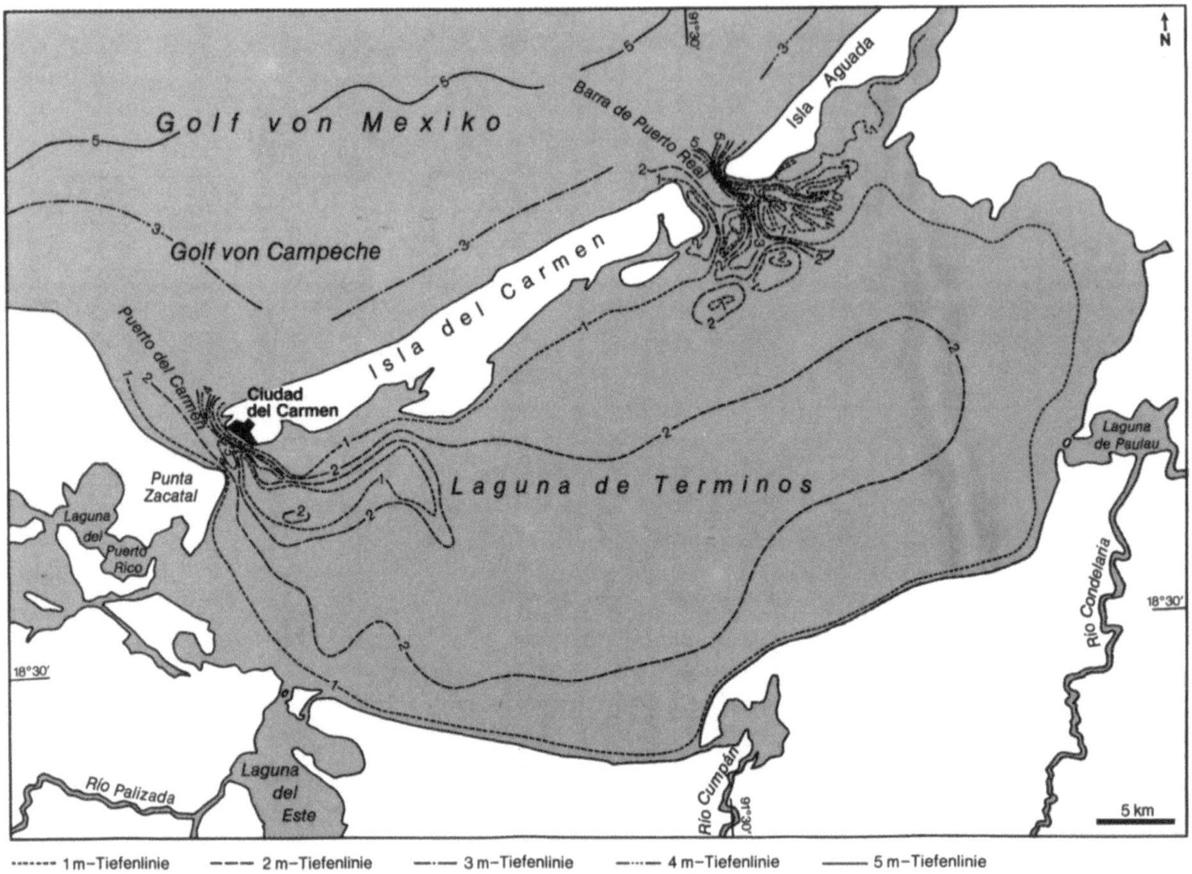

Laguna de Terminos, Campeche Bay and Gulf of Mexico

Chart of the area of the satellite photograph
scale 1 : 4 500 000,
with bathymetric topography, depth in m
based on A. Yañez, 1963

Water mass interaction, lagoonal-oceanic, visible due to sediment laden waters

Change detection of processes on diurnal time scale and comparison of Gemini photograph from 1965 with Skylab photograph from 1965 of this area, demonstrating the long term significance of transient processes

Laguna de Terminos and Campeche Bay, Gulf of Mexico

NASA colour photograph, Gemini mission 5, Mag. 1, frame 31; date 22. Aug. 1965.
Laguna de Terminos and Campeche Bay on the peninsula of Yucatán, Mexico; hand operated camera;
The photograph is a slightly oblique one; geographic distortion is small. The scale of this photograph is approximately 1 : 600 000 (picture scale). The azimuth of the sun is westerly. The sun's elevation is high, about 65 degrees. This exposure was made during ebb tide when the sediment laden waters were pouring out of the Laguna's west channel into Campeche Bay.

Visibility of a tide caused pumping system of a lagoon in interaction with the ocean due to the colour and structure of sediment laden water masses on the satellite photograph near the shrimp fishery of Campeche Bank. Sediment acts as a natural tracer. The longshore current from the northeast carries the turbid waters westward along the coast several miles out into the Bay.
Among the earliest specific scientific applications of orbital photography was the use of Gemini photography, reported on the International Sedimentological Congress in Great Britain (Wobber, 1967).
This system is of interest because the function of the eco-system of this lagoon depends on the hydrographic system of water mass exchange. There are oyster banks in the lagoon and there is shrimp fishery in the off-shore area, both of economical importance. The life cycle of shrimps is such that larvae drift into the lagoon and the adults out of the lagoon to deposit their eggs in the offshore area.

Description of the area

The Laguna de Terminos, also called: "Laguna del Carmen", is situated on the coast of the Gulf of Mexico in the state of Campeche, Mexico. The lagoon is about 60 km long and 25 km wide. It is shallow; the water depths are up to 2 m and in some parts somewhat more. The area of the lagoon extends to about 1200 square kilometres.
The 5 m depth contour runs at a distance of 2 to 5 km in front of the coast. The 10 m depth contour runs at a distance of 5 to 15 km in front of the coast, and even at a distance of 25 km from the coast the water depth is only 10–20 m.
The greater part of the lagoon is separated from the ocean by the barrier island Isla del Carmen. To the east and the west of the Isla del Carmen the lagoon has two natural channel connections to the sea, i. e. to the Gulf of Mexico. The eastern channel is called Canal Barra de Puerto Real. The western channel is called Canal Barra del Carmen.

"Isla del Carmen is not a single sand ridge, nor is it composed of parallel beach ridges running the entire length. Rather, it is three separate barrier islands that have been fused together by the development of a narrow sand bar, a dune ridge, and mangrove swamps, each island having a fanning set of beach ridges flaring toward the west." Thom, B. G., 1967.

The Barra de Puerto Real has less than 4 m water depth at its entrance. A flat sand bank on parts of which breakers sometimes occur when strong winds prevail, extends from Punta del Tigre for about 2.5 nautical miles in a WNW direction. One nautical mile west – and two nautical miles NNW – from the same spit there are two shoals with depths of only about 1 m. There are several shoals about 1.25 nautical miles WNW of the north-eastern foreland of the Isla del Carmen.
Canal del Carmen. Barra Principal is the deeper of the two entrances to Laguna de Terminos with depths in excess of 7 m (Ayala, 1963). The entrance lies between the west spit Punta Atalaya on the Isla del Carmen and the Punta del Xicalango on the mainland.
From Punta del Xicalango a flat bank extends about 3 nautical miles to the north and northeast. Between Punta del Xicalango and Punta del Zacatal, a bar extends in front of the coast about 1.5 nautical miles in width. Within the entrance between Ciudad del Carmen and Punta del Zacatal lies the shoal Bajo de los Bagres.
A navigable waterway, at least 4 m deep, passes between the shoals on both sides of the entrance. In certain small areas there are troughs up to 10 m in depth. In the west, a row of smaller lagoons and river estuaries, altogether about 30 km in length, joins the Laguna de Terminos.

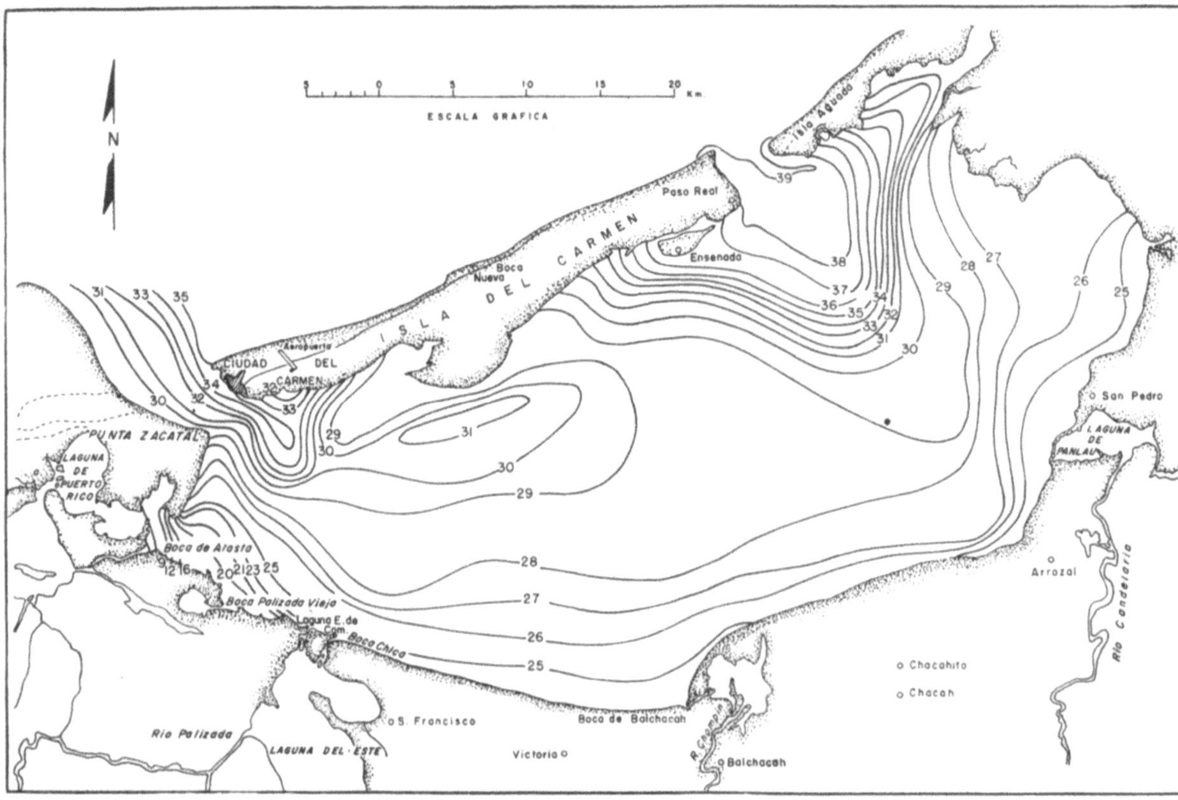

Distribution of salinity of bottom water, in parts per thousand
for March and April 1959, Laguna del Carmen
from A. Yañez, 1963, Fig. 16, p. 30.

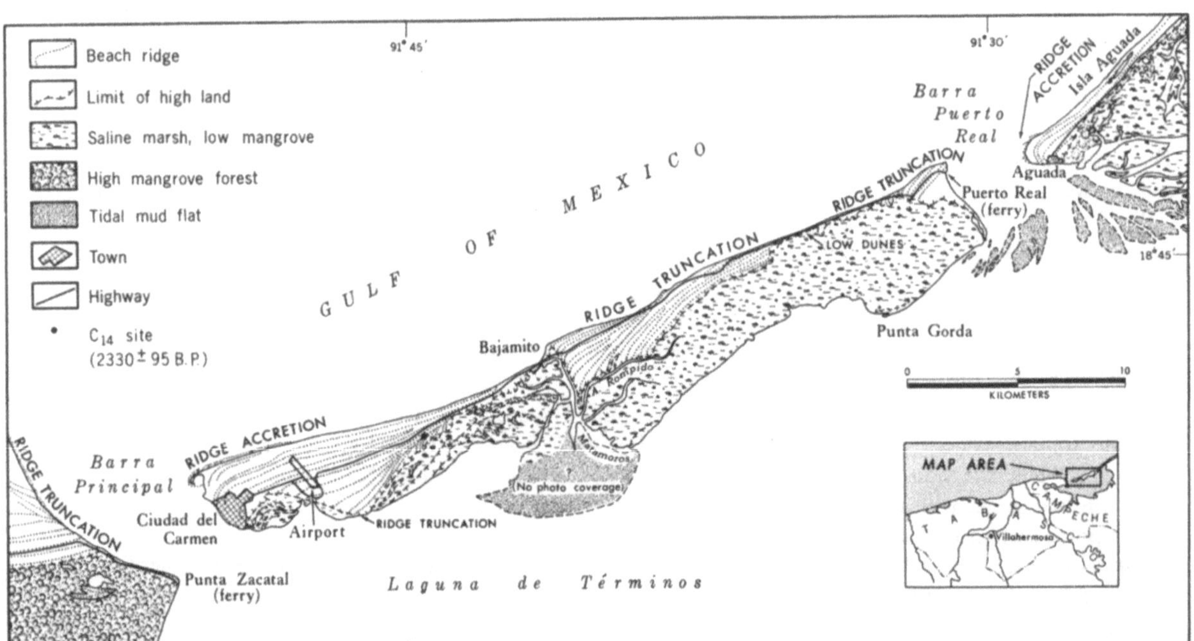

Location and geomorphology of Isla del Carmen, Campeche, Mexico. Comparison with Skylab photograph gives the result, that some of the features of coastal environments as drawn on this map can be detected and identified on the Skylab photograph. No particular study has been made in this publication since this section deals with streams and currents.
Thom, B. G., 1967, Problems of the development of Isla del Carmen, Campeche, Mexico: Z. f. Geomorph. vol. 13, pp. 406–413.

Morphology

The bottom of the Laguna de Terminos is flat, and, except near the entrances, without channels or banks. A bar has however developed inwards into the lagoon in the Canal de Puerto Real. It is fanshaped 5 to 10 km in diameter, and is composed of finger-shaped sandbars and channels (like a submarine delta). Also in the western entrance, Canal del Carmen, a bar extends seawards into the Gulf of Mexico, two deeper main channels landwards into the lagoon and a flat bank exists in the lagoon. The barrier island Isla del Carmen is composed of sandy beach ridges. Towards the lagoon a muddy strip is connected to it.

Hydrography: rivers

The four larger rivers which discharge fresh water into the Laguna de Terminos, are from west to east:

1. Rio San Pedro out of the system of the lower Rio Usumacinta into the estuaries and small lagoons at the western end of the Laguna de Terminos
2. Rio Palizada out of the system of the Rio Usumacinta into the estuary Laguna del Este
3. Rio Chumpán also called Cumpán
4. Rio Candelaria in the eastern part of the lagoon.

The drainage area of the Rio Candelaria is located on the peninsula of Yucatán, the other rivers come from the Gulf Coast plains, or the hinterland. Some rivers are widened like estuaries and bays at their entrance to the lagoon.

Tides and currents

The main flood tide stream enters the eastern channel Canal de Puerto Real; the ebb tide stream flows out of the westerly channel Canal del Carmen. These main tidal streams cause the position of the bars.
The tidal streams are described as follows: (DHI Nr. 2050, translated). In the eastern entrance the incoming flood tide stream rates 1 knot and flows SE-wards beyond the bar and continues further within the deep channel. The outgoing ebb tide stream is intensified by the water masses of the rivers which enter into the lagoon, and rates up to 3 knots. When the incoming flood tide stream at the western entrance is increased by the sea breeze it rates up to 1.5 knots. If the outgoing tide is intensified in late evenings and in early mornings by the land breeze, it rates up to 2 knots. The tidal range amounts to 0,5 m.

Salinity

The inflow and outflow of water from the Gulf of Mexico is evidenced by the distribution of salinity in the Laguna de Terminos. According to Yañez, Fig. 15 and Fig. 16, saltwater, with 39 ‰, passes into the eastern channel and distributes in the lagoon in the eastern part fairly evenly in the whole water column from the bottom to the surface, so that in the southern inner edge of the lagoon 25 ‰ salinity still occurs. In the western part of the lagoon water up to 35 ‰ salinity enters through the Canal del Carmen, but mainly only near the bottom and not far into the lagoon. The rivers discharge fresh water into the lagoon so that water of different salinity is mixed up. (Compare Map of Yañez).
The effect is an electrolytic coagulation of the colloidal particles suspended in the water masses to a sedimentation process.

Wind and weather

The currents of the lagoon are also influenced by winds. Ciudad del Carmen is situated in the region of the NE-trade winds. This system is sometimes, from November to April, interrupted by Northers. From June to October there often is calm and rainy weather (rainy season). Sea- and land breezes are diurnal.

The transition time from land to sea breeze, 10.00 to 14.00 is usually calm, until the sea wind sets in from north or northeast.
The wind direction is more important for the currents in the middle region of the lagoon than the tidal currents.
The surface coastal water currents flow on the average, seasonally in February from west to east, in May from east to west; in August from southeast to northwest (season when the photograph was taken).

Sediments

Amado Yañez, 1963, gives a complete report on sediments and sedimentation of the Laguna de Terminos.

On the satellite photograph little is recognizable on the bottom of the lagoon, because transparency is poor (less than 1 m), and the waters are laden with sediments.
Sand drifts from the ocean (Gulf of Mexico) into the Laguna de Terminos through the eastern channel with the tide and wind driven water masses. A bar, the "Barra", is situated in the lagoon at the south end of the channel. The rivers, depending on their drainage areas, discharge different types of sediments into the lagoon.
There are several chemical processes of inorganic and organic cycles going on in the lagoon, e. g. there are extensive oyster banks present in front of the river mouths and estuaries in the lagoon which are important for the carbonate cycle.

Interpretation of the satellite photograph

Water masses and currents and change detection of processes of diurnal time scale

This satellite photograph was taken during ebb tide when the sediment laden waters were pouring out of the west channel of the lagoon into Campeche Bay. The longshore current from the northeast carries the turbid waters westward along the coast several miles to seaward of the Bay. Tonal variation clearly shows the turbid discharge of water from Laguna de Terminos into the Bay of Campeche. Sediment laden water masses are identifiable in the Laguna de Terminos; two currents are identifiable:
in the Canal del Carmen and west of this channel by the whitish colour (tonal variation) and the striped-like bunching (pattern); the channel of the ebb tide stream is clearly marked.
Even though the circulation of the water masses in the lagoon system is known in general, it is now possible to distinguish the boundaries of the turbid water masses in the Gulf of Mexico with the help of the interpretation of the satellite image. (Compare map and photograph).
It is a component of the tidal pump-system which, even though the tidal range is small, shows the interaction of the water masses between the lagoon and the Gulf. The information from the satellite photograph shows how far these outflowing sediment laden water masses reach into the Gulf. The coastal current borders the water masses, coming out of the Canal del Carmen to drift near to the western beach.
A current system far away from the coast is identifiable on the satellite photograph: a light blue water mass which is discernable as a broad band (about 20 km wide) at a distance of up to 30 km in front of the coast, north of Isla del Carmen. At its northern edge, towards the water masses of the Gulf, eddies are identifiable, which appear dark blue. This water mass has drifted away from the coast with the ebb tide current.
In addition there are two more phenomena identifiable in the water. In the southeast corner of the Laguna de Terminos water masses of the Rio Candelaria are identifiable as a dark colour in front of its mouth.
Along the coast, east of the Barra del Carmen, in front of the beach, a 2 km band is identifiable, whose structure shows sandy reefs of the foreshore zone.
Bottom topography in the surf zone is produced by driving forces such as waves and currents.
The general transparency of the water in the photograph is small so that bottom topography below low water level can hardly be recognized (only in some channels in the Canal del Carmen). However, the phenomenon of sediment clouds in the water enables one to detect in an overview the currents of the water masses and the boundaries of their areas, i. e. concerning the lagoon, the area of interchange of lagoonal and marine waters.

Terrestrial features and inter-tidal zone

For terrestrial features of the coastal environment, compare the interpretation map.
For features of the inter-tidal zone, compare the interpretation map and the comparison of this Gemini photograph with the Skylab photograph demonstrating long term significance of transient processes.

Comparison of orbital remote sensing products of NASA, Gemini satellite photograph of 1965 and Skylab satellite photograph of 1973, example of evaluation of orbital remote sensing products as documents demonstrating the long term significance of transient processes (photograph taken nearly eight years apart) and change detection of processes of diurnal time scale

(Published in Skylab Earth Resources Data Catalogue, NASA, 1974, pp. 108). In this publication the pictures are presented in black and white copies of the Gemini and Skylab photographs. Compare Gemini photograph with colour photograph of this publication.

Laguna de Terminos and Campeche Bay, Gulf of Mexico

Photograph on the left:
Gemini 5 mission, no. S 65-45-765
from 22. Aug. 1965
viewing mode: slightly oblique

Photograph on the right:
Skylab 2 mission, no. SL 2-04-307
from Aug. 1973
viewing mode: vertical
taken with the EREP multispectral camera
with better transparency

Coastal and estuary circulation pattern can be observed in the same way, as shown in both satellite photographs, Gemini (left) and Skylab (right).
Note that the wind direction, as deducted from cloud pattern, is identical in the two photographs.
Note cloud distribution and pattern as land – sea breeze development (i. e. change detection).

On Gemini photograph:

Sediment is obviously being discharged at the left-hand or western end of the Laguna, and a delta can be seen on the inside of the eastern opening.

The sandbar at the approaches of the western entrance to Laguna de Terminos is visible (marked as "G 1" on the Gemini photograph).

The sediment laden waters of the Rio Palizada are discernible but not developed as plume-like pattern and not with sharp boundaries.

Cloud distribution and pattern as phenomenon of full developed sea – land breeze system at noontime (local observation time 14.00)

The Gemini photograph was studied separately in considerable detail in this publication (compare colour photograph and interpretation map)

On Skylab photograph:

In 1973 it clearly shows that the sediment transport mechanism is unchanged and that the eastern delta is expanding. Better separation of terrestrical-intertidal features as covered or uncovered due to multispectral photography.

The sandbar at the approaches of the western entrance to Laguna de Terminos is not well visible on the Skylab photograph.

The sediment laden waters of the Rio Palizada are discernible as plume-like pattern with sharp boundaries (marked as "SL 1"). Better separation of water masses due to multispectral photography.

Cloud distribution and pattern as phenomenon of sea – land breeze in development at late morning time (local observation time 9.00).

The Skylab photograph was studied separately in considerable detail: Earth resources surveys from spacecraft, prepared by U.S. Army Corps of Engineers for Earth Resources Survey Program, Space Applications Programs Office, NASA, vol. II, no date, p. 6 ff (N. P. Psuty and J. S. Bailey).

"If examination of these pictures only results in a logical deduction of such simple facts that sediment always goes one place and never goes another, that information can be of great value in planning human activities.

The fact that the same sediment transport mechanism is continuing today as it was eight years earlier strongly suggests that this circulation pattern would be of crucial interest to planners of any construction activity along this part of the coast or in the Laguna, including projects involving harbour and shoreline maintenance."
NASA, Skylab catalogue, 1974, pp. 108.

Bibliography

Amado–Yañez, C., 1963, Batimetria, salinidad, temperatura y distribucion de los sedimentos recientes de la Laguna de Términos, Campeche, México: UNAM, Inst. Geol., Bull. no. 67, part 1, p. 47, maps. México D. F.

Ayala–Castañares, A., 1963, Sistemática y distribucion de los foraminíferos recientes de la Laguna de Términos, Campeche, México: UNAM, Inst. Geol., Bull. no. 67, part 3, 130 p., 71 fig., México D. F.

García–Cubas, A. jr., 1963, Sistemática y distribucion de los micromoluscos recientes de la Laguna de Términos, Campeche, México: UNAM, Inst. Geol., Bull. no. 67, part 4, 55 p., 28 fig., México D. F.

Psuty, H. P., 1965, Beach-ridge development in Tabasco, Mexico: *Ann. Ass. Am. Geogr.*, vol. 55, p. 112–124.

Silva–Barcenas, A., 1963, Sistematica y distribucion de los géneros de diatomeas de la Laguna de Términos, Campeche, Mexico: UNAM, Inst. Geol., Bull. no. 67, part 2, 31 p., 12 fig., Mexico D. F.

Tamayo, J. L., 1946, Datos para la hidrologia de la Republica Mexicana: Inst. Panamericano Geogr. Hist., publ. 84, 2 vol.

Thom, B. G., 1967, Mangrove ecology and deltaic geomorphology, Tabasco, Mexico: *J. Ecology*, vol. 55, p. 301–343.

Thom, B. G., 1969, Problems of the development of Isla del Carmen, Campeche, Mexico: *Zeitschr. Geomorph.*, vol. 13, p. 406–413.

West, R. C., N.P. Psuty and B. G. Thom, 1969, The Tabasco lowlands of Southeastern Mexico: Coastal Stud. Inst. Baton Rouge, Techn. Rep. no. 70, 193 p.

Maps

Chart ONC – WAC code J – 25

Nautical chart, Deutsche Seekarte no. 530, DHI, in scale 1 : 1 000 000, Golf von Mexico, südwestlicher Teil von Cabo Catoche bis Coatzacoalcos, 4. August 1964. (No soundings are shown in the lagoon).

Nautical chart, mapa no. F. H. 500, in scale 1 : 909 660, Tampico a Progreso: Secretaría de Marina, E. U. Mexicanos. (No soundings are shown in the lagoon).

Mapa Estados Unidos Mexicanos, in scale 1 : 500 000, sheet Campeche 15 Q-VI 1957. (No soundings are shown in the lagoon).

German-Pilot: (West Indies Handbook) „Westindien Handbuch III. Teil", Nr. 2050, DHI, Hamburg, Ausg. 1974.

Visualising the „respiration of a lagoon“, due to detectable turbid waters, caused by the tidal pumping system, hydrographic processes and some submarine topography – presented with satellite photography

Pamlico Sound lagoon and coast and offshore of the Atlantic Ocean of North Carolina and Virginia, USA, around Cape Hatteras

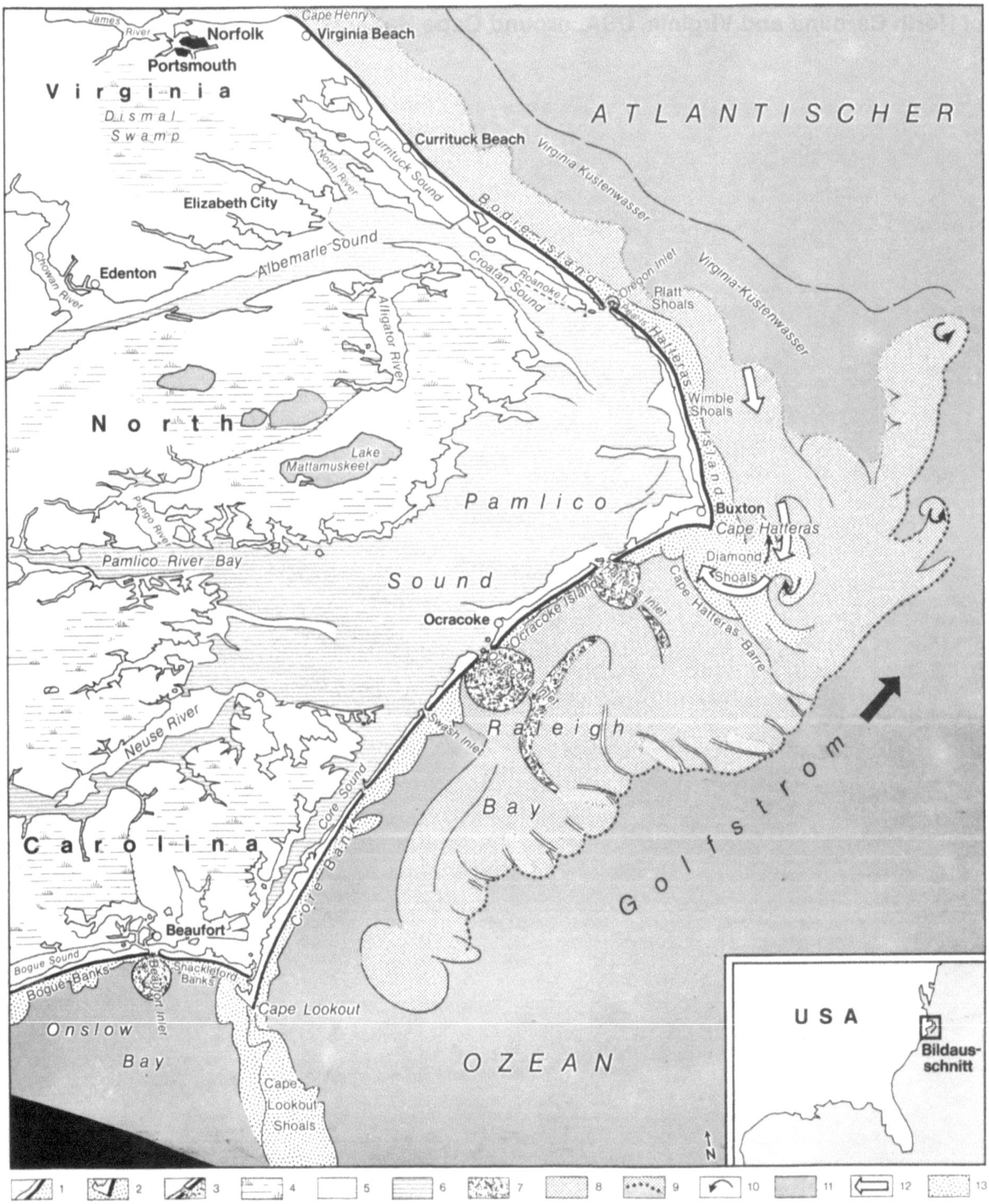

Pamlico Sound Lagoon

Interpretation map, distortion as on satellite photograph
scale in the southern part approximately 1 : 1 000 000
scale in the northern part approximately 1 : 1 800 000

Pamlico Sound Lagoon and coast of the Atlantic Ocean around Cape Hatteras, North Carolina and Virginia, USA

Interpretation map, from NASA satellite colour photograph

APOLLO mission 9, no. E 20–3128

Interpreter H. G. Gierloff-Emden, 1975

Legend of topographic and hydrographic features classified in zones with selected discriminated objects (generalized: no differentiation of distinct water levels on this small-scale map)
Scale varies from 1 : 1 000 000 to 1 : 1 800 000 (caused by oblique photograph)

I above High Water level: terrestrial

mainland on the map as white area

1 barrier island and beach

2 spits

(4) mainland, lowland hardwood forest, marshes and (compare II) salt marshes

II between High and Low Water level: inter-tidal zone

3 inlets

4 salt marshes (and marshes)

III below Low Water level: lagoonal and marine

(3) compare II

5 waters with no turbidity – mostly of river estuaries

6 lagoonal waters, very turbid

7 plumes of lagoonal waters (plume pattern), flowing out of lagoon, caused due to the ebb tidal stream into the Atlantic Ocean (representing flow dynamics)

8 furrows of moderately turbid water in the Atlantic Ocean, partly covering shoals (overlies partly 13)

9 boundary between turbid lagoonal and coastal waters and Gulf Stream waters in the Atlantic Ocean

10 eddies

11 oceanic water

12 Virginian coastal waters from north, surrounding Cape Hatteras and overflowing shoals

13 submarine bars, sandwave fields, and shoals in the foreshore (partly superimposed from 8) and offshore region

14 Gulf Stream: on the map as "Golfstrom" with black arrow

Satellite photograph:

Pamlico Sound Lagoon and Coast of the Atlantic Ocean, around Cape Hatteras
North Carolina and Virginia, USA

Flight information data:

Mission: *NASA – APOLLO 9*

Date: *12. March 1969*

Photograph: *colour*

Magazine: *E 20   frame no. 3128*

Camera: *Hasselblad 500 C*
70 mm photography

Lens: *Zeiss Planar f/2,8; 80 mm*

Film: *Kodak Ektachrome*

Filter: *haze*

Altitude above mean sea level: *116 naut. miles*

Geographical position of the principal point:
latitude: *35 degrees 04 min. N*
longitude: *75 degrees 40 min. W*

Chart: *ONC – WAC code G – 21*

Local observation time: *10.15*

Sun azimuth: *95° to 115°*

Sun elevation: *40°*

Photograph viewing mode: *oblique exposure axis to 0°*

Cloud cover: *30 % alto cumulus*

Scale of the printed picture: *southern part 1 : 1 000 000 (approximate) northern part 1: 1800 000*

Colours of the printed picture (Munsell scale):
The colours of the printed photograph occur more yellowish and greenish yellow in comparison to the colours of the NASA copy, which shows up less yellowish, but more bluish and purplish pink;
result: lagoonal waters, very turbid, (6) of the interpretation map occur on printed photograph yellowish, on NASA copy greenish.

NASA – APOLLO 9

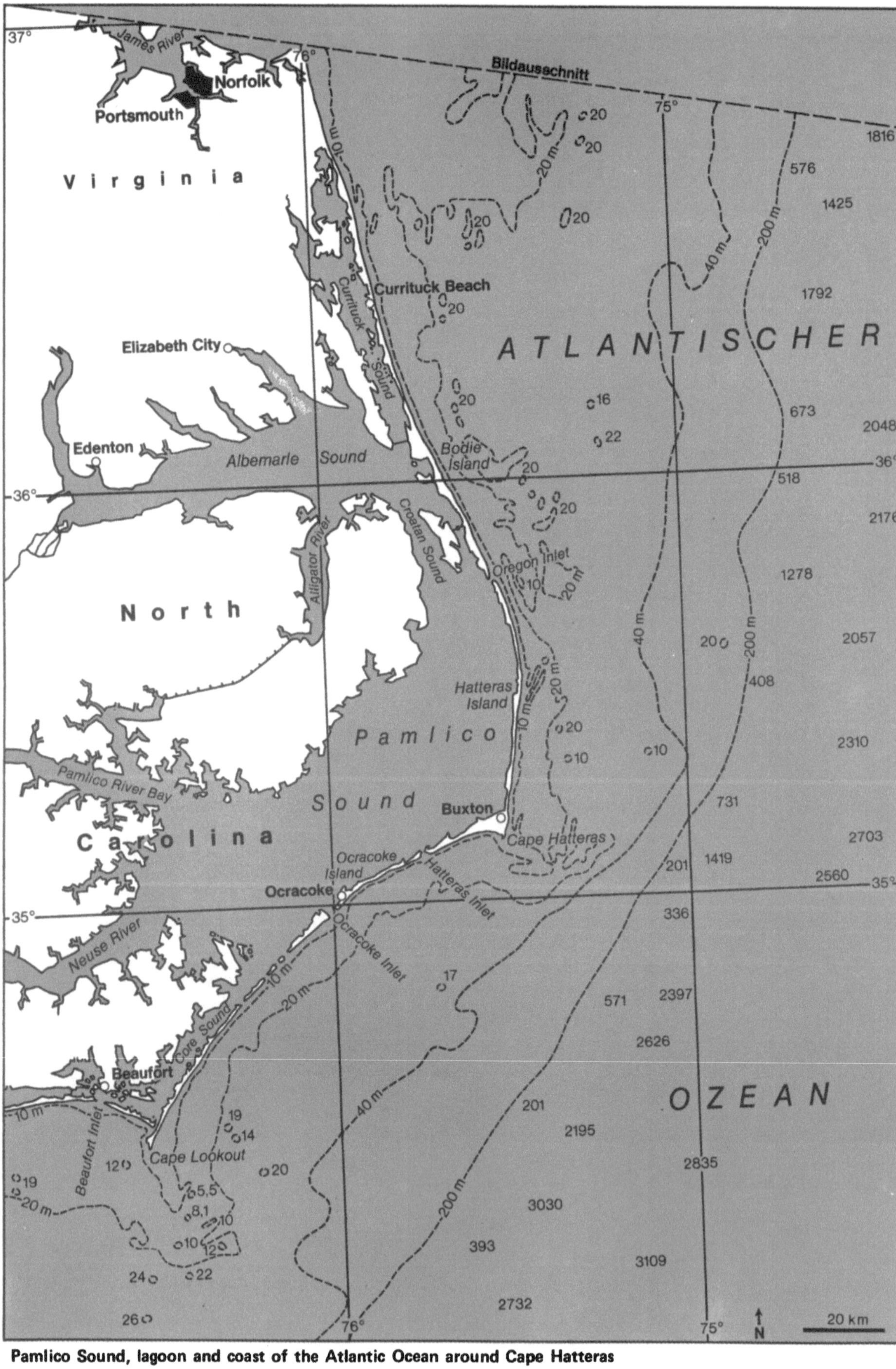

Pamlico Sound, lagoon and coast of the Atlantic Ocean around Cape Hatteras

Chart of the area of the satellite photograph scale approximate 1 : 1 450 000.
bathymetric contourlines, based on nautical charts

Visualising the „respiration of a lagoon“, due to detectable turbid waters, caused by the tidal pumping system, hydrographic processes and some submarine topography – presented with satellite photography

Pamlico Sound lagoon and coast and offshore of the Atlantic Ocean of North Carolina and Virginia, USA, around Cape Hatteras

NASA colour photograph, Apollo mission 9, 69-HC-343, AS 9-20-3128; date 12. March 1969; taken by astronauts Mc Divitt, Scott and Schweickart; hand operated camera; altitude 116 nautical miles.
The photograph is not a vertical one; it is an oblique photograph. Direction of the axis of the camera is to the north, distortion mainly in the northern part of the photograph; scale in the southern part approximately 1 : 1 000 000, in the northern part approximately 1 : 1 800 000. The photograph was taken in the morning time at 10.15, the azimuth of the sun is easterly, the elevation of the sun is 40°.

Description of the area

The photo shows a lagoon area with flats and marshes and sandy barrier islands, which appear in the picture as a yellow-grey line jutting into the sea in spits at Cape Hatteras and Cape Lookout. The extensive tidal lagoon to the left, west of the spits is Pamlico Sound. The distance between the two capes is actually about 150 km. The Pamlico – Albemarle Sound covers 6 630 square kilometres.
Most parts of the lagoon of Pamlico Sound are very shallow, depths are no more than 5 m; mean depth is 3,6 m. The tidal range near Cape Hatteras is 2 m. The tidal current runs at 2 knots.
"The mainland beach and associated spits are nourished from the eroding headland, but the barrier islands, having originated as mainland beaches detached during the post-Pleistocene sea-level rise, now appear to be fed by shelf sands." D. J. P. Swift.
In the lagoon area enclosed behind the spits are partly dried out flats, and here and there the grey-brown coloured water-covered flats or shallows, can be seen.

Hydrographic processes: ebb-stream plumes from the lagoon and dissipation of such discoloured water masses

The most prominent features of this satellite photograph are the plume-like patterns of water masses, pouring out of the inlets of the lagoon, (at Ocracoke, Hatteras and Beaufort inlets) and the furrows further from the coast, with sharp boundaries to the blue coloured ocean waters. The "plume patterns" of turbid waters have to be described as an ebb-flowing current, travelling as jets.
R. L. Mairs from the U.S. Naval Oceanographic Office carried out an intensive study of the hydrographic phenomena in this picture, published 1970 in "Photogrammetric Engineering", with the title "Oceanographic Interpretation of Apollo Photographs, coastal oceanographic and sedimentologic interpretation of Apollo 9 space photographs; Carolina's continental shelf, USA".
Following Mairs, the following explanations are given:
The lagoon Pamlico Sound is covered with sediments of mostly fine sands and silt, with only a small amount of organic matter. The plume pattern turbid waters, pouring out of the inlets, represent ebbstream flowing currents. The turbid water masses are caused by wind generated shallow water turbulence in the lagoon, rather more than by discharge of the rivers which flow into Pamlico Sound; major rivers are the Pamlico river and the Neuse river (moderate rainfall was reported some days before but only small amounts of sediments were discharged). These plumes of turbid water are opaque and have no transparency and cover in particular the submarine topography outside the inlets. On the lagoonal side of the inlets lie delta-shaped sand bars caused by the sand transport of ingoing flood water masses. The photo was taken 5 hours after the start of the ebb-stream, (which reached a maximum velocity at 20.00) and slack-water was expected two hours after the photo was taken.
The water mass, represented by the plumes, travels as an undiluted jet with an average current velocity of

1,6 knots outside the inlets (calculated by Mairs), so that a distance of 7,7 nautical miles could have been reached since leaving the inlet. A check on the photo showed that it was 6,7 nautical miles. Such a water mass of lagoonal character can travel a long distance before mixing with oceanic waters. Many studies have been made on this subject, as for instance that published by I. V. Samojlow, 1956, "Die Flussmündungen", who points out that such water masses, pouring out of an estuary or a lagoon into the ocean, are stabilized gyroscopic like a jet, and often extend to 4 to 5 times as far as the channel is wide.

"The Ocracoke Inlet plume extends out over the shelf between four and five inlet widths. Once beyond the limit of four channel widths, lateral turbulence predominates." R. L. Mairs.

Although with this satellite photo it is not possible to detect which processes occur in the deeper layers of the water masses, it is known that this estuarial ebb water of Pamlico Sound remains for a long time a surface flow and layer, due to its lower density and salinity in comparison with the ocean water.

"The plume that flows out the inlet during ebb flow does not flow back during flood flow. Some reversal of flow undoubtedly takes place near the inlet mouth, but not enough to destroy the symmetry of the outflow". R. L. Mairs.

The Pamlico Sound water is different from Carolina shelf water and from Virginia coastal water, concerning the two parameters suspended sediment and salinity: Pamlico Sound water has much more suspended sediment (2 mg/l) than the others, which have 0 to 1 mg/l.

B. A. Buss and K. S. Rodolfo, 1972, "Suspended sediments in continental shelf waters off Cape Hatteras, North Carolina": in: D. J. P. Swift, 1972, "Shelf sediment transport".

Plumes and sediment transport, flow dynamics discriminated by remote sensing techniques

The turbid sediment plume of Ocracoke Inlet maintains a symmetrical shape outside the lagoon near the coast, since winds and currents do not destroy its configuration. Further from the coast in the deeper ocean, the plumes are deformed by the actions of the winds and currents.

Beyond the symmetrically shaped plumes, there are the remnants of the previous ebb flow; they are deformed and diluted with ocean waters, so they appear to be less sediment laden, as their deeper blue grey colour indicates. Mairs explains that the sinking of these waters is caused by cooling.

The south-western boundary of those plumes is curved, due to shelf form and coriolis force. The outer edge of the water masses has the appearance of furrows, due to contact with the Gulf Stream. This contact is sharper in the southern part of the region (compare photograph).

Further offshore, there are some frictional eddies over the shallow shelf edge. The vortex motion is subject to change detection with satellite imagery. The Gulf Stream flows at a distance of 22 miles offshore from Cape Lookout and 20 miles offshore from Cape Hatteras. The axis of the Gulf Stream is located in depths of 180 m to 1800 m over the continental slope.

The line where the blue Gulf Stream water starts is very marked and precisely coincidental with the 7° C isotherm. D. F. Bumpus, 1973, reported on his drift-bottle experiments on current phenomena in the region offshore of the U. S. Atlantic Coast. H. G. Stumpf, 1974, reported on studies with NOAA-2 satellite photographs. He could identify the boundary of the Gulf Stream due to interpretation of infrared images.

Seasonal change in the current system in this area is of importance, but is not discussed here.

R. L. Mairs worked out a theoretical model for the movement of watermasses in this area and for the ocean current system. This is not the object of the interpretation referred to here.

There is one more hydrographic phenomenon which can be detected on the satellite photograph: coastal water masses from the north (the Virginian coast), drive southward around Cape Hatteras; these water masses are sediment laden and visible on the photograph, but because they cover the shoals off Cape Hatteras, the information from the satellite photo is ambiguous. Both phenomena – sediment laden waters and shoals – cause similar water discolouration.

This phenomenon could be explained only by means of pre-information, i. e. the personal knowledge of the authors El Ashry and Wanlass, not by interpretation of the satellite photograph, which contains ambiguous information in that case.

Submarine topography

Southward of Cape Lookout, the shoals extend far into the ocean, about 18 nautical miles offshore, as shown on the nautical charts of the area. These shoals have a winding configuration. So the above mentioned pre-information is essential if the discolouration in this region is to be identified as a phenomenon of water transparency, representing the submarine bottom topography. The transparency of the water along the Atlantic coast of this region varies from 2 to 35 m (as measured by Secchi disc.). Nevertheless, some water discolouration may be caused by turbulence in this area. F. P. Shepard and H. R. Wanless, 1971, in their book "Our Changing Coastline", which is a classic in its field, give a chapter with aerial photographs of Cape Hatteras and Cape Lookout spits, underwater sand bars and shoals and their coastal and submarine processes and geomorphology.

Clearly it points out the overlapping of sedimentation and crosswise erosion of shoreline ridges and submarine bars, formed alternately from north-east or south-west by currents and wave actions. The winding pattern of the bars and shoals is an effect of those processes.

The morphology of the cape associated shoals and sediment transport in this area is referred to in the study from D. J. P. Swift and al., 1972: "Holocene Evolution of the shelf surface, Central and Southern Atlantic Shelf of North America.

R. Dolan, 1973, reports on the phenomenon of crescentic sandbars along beaches of the barrier islands and sandwave fields as migrating sandmasses near Cape Hatteras, which range to 1 km in length.

The submarine topography of the barrier entrances cannot be observed visually as it is covered by turbid waters.

D. W. Folger reports that sand sediments from the open sea are transported as washover fans inside to the barrier islands of the lagoon.

Important for interpretation of the satellite photograph is:

The light grey-brown tone that is also distinguishable off Capes Hatteras and Lookout. Here lie so-called shoals, flat underwater sandbanks whose cloud-like form can be very clearly distinguished on large-scale maps. The startling turquoise colour of the area seawards from Cape Hatteras can be explained either as shallows or as suspended sediment. Local conditions are such that there is a possibility that both phenomena exist. The pictorial information is ambiguous.

Only the semicircular patch near the gap in the spit between the Capes can with certainty be identified as sedimentary waters. A special study on these problems has been published:

M. T. El Ashry and H. R. Wanless, 1968: "Photo interpretation of shoreline changes between Capes Hatteras and Fear (North Carolina)": Marine geology 6, pp. 347–379.

Bibliography

El Ashry, M. T. and H. R. Wanless, 1968, Photo interpretation of shoreline changes between Cape Hatteras and Fear, North Carolina: *Marine Geol.*, vol. 6, no. 5, p. 347–379.

Bumpus, D. F., 1955, The circulation over the continental shelf south of Cape Hatteras: *Trans. Amer. Geophys. Union*, vol. 36, p. 601–611.

Bumpus, D. F., 1973, A description of the circulation on the continental shelf of the East Coast of the United States: in B. A. Warren (ed.), *Progress in oceanography*, vol. 6, p. 111–158.

Climatological data, North Carolina, March 1962: ESSA, vol. 74, no. 3.

Dolan, R., 1973, Coastal processes, high-altitude aerial photographs aid in investigations: *Photogramm. Engineer.*, no. 3, p. 255–260.

Dolan, R., Beach changes on the outer banks of North Carolina.

Folger, D. W., 1972, Texture and organic carbon content of bottom sediments in some estuaries of the United States: *Geol. Soc. America*, Mem. 133, p. 391 ff.

Haight, F. J., 1942, Coastal currents along the Atlantic Coast of the United States: Coast and Geodetic Survey, Spec. Publ., no. 230.

Harrison, W., J. J. Norcross, N. A. Pore and E. M. Stanley, 1967, The circulation of shelf waters off the Chesapeake Bight: ESSA Profess. Pap. no. 3.

Haupt, I., 1972, Fernerkundung der Erdoberfläche durch Satelliten: Beilage zur Berliner Wetterkarte 122/72, SO 37/72, p. 22, fig. 10 Temperaturverhältnisse im Bereich des Golfstromes.

Mairs, R. L., 1970, Oceanographic interpretation of Apollo photographs, coastal oceanographic and sedimentologic interpretation of Apollo 9 space photographs. Carolina's continental shelf, USA: *Photogramm. Engineer.*, vol. 36, no. 10.

Oostdam, B. L. and R. R. Jordan, 1972, Suspended sediment transport in Delaware Bay: *Geol. Soc. America*, Mem. 133, p. 143.

Shepard, F. P. and H. Wanless, 1971, Our changing coastlines: McGraw Hill, 579 p., especially p. 104–117.

Stumpf, H. G., 1974, A satellite-derived experimental Gulf Stream analysis: *Marine Weather Log*, vol. 18 (May), p. 149–152.

Swift, D. J. P., D. B. Duane and O. H. Pilkey (eds.), 1972, Shelf sediment transport, process and pattern: Dowden, Hutchinson and Ross, Shroudsbury, Pennsylvania, 656 p.

Tidal current tables, 1969, Atlantic Coast of North America: ESSA.

Maps

Chart ONC – WAC code G – 21.

Nautical chart, Deutsche Seekarte no. 450 and 451, DHI, in scale 1 : 500 000, Sandy Hook bis Kap Hatteras, Kap Hatteras bis Savannah, 1971.

Nautical chart, Deutsche Seekarte no. 458, DHI, in scale 1 : 200 000, Kap Henry bis Kap Hatteras.

Interaction of lagoonal waters with oceanic waters – complicated due to oceanic phenomena and processes change detection

Lagoons and Coast of Sonora, Mexico, around Guaymas and the middle part of the Gulf of California

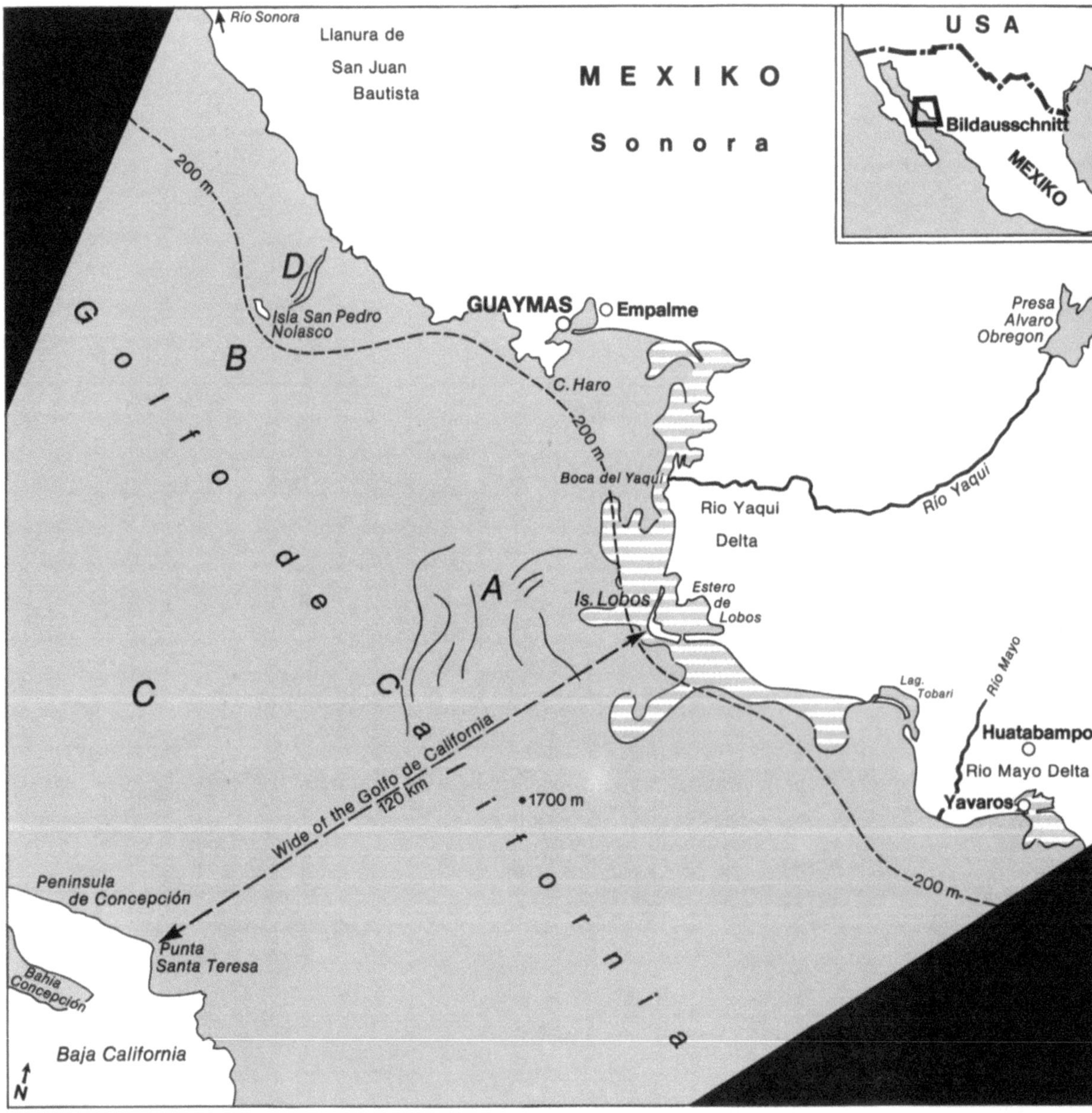

distribution of lagoonal water masses (estero = lagoon)

Coast with lagoons of Sonora, Mexico around Guaymas, and Gulf of California

Sketch within frame of facing NASA satellite colour photograph

APOLLO mission 7, no. 5–1632

Legend of topographic and hydrographic features with selected discriminated objects

Scale varies due to oblique photograph

Submarine topographic features in shallow water regions of the foreshore region and especially of the coastal lagoons of Sonora, and hydrographic features as distribution of outflowing lagoonal water masses in the Gulf of California can be identified on the photograph.

A sinuous tongues of discoloured water, south of Guaymas, interpreted to be phytoplankton blooms, red tide (biological activity) indicating streamline features as slicks on the surface of the sea

B rough sea state and cloud lines

C calm sea state; sea-surface state may be determined from the photographs by means of tone and texture changes which are responses to areas of rough and calm water; details are not so clearly oceanic in nature nor are they definitely of atmospheric character

D probable standing waves near Isla San Pedro Nolasco were recognized which are attributed to either the effects of local current circulation or strong north – south movements of water in the Gulf of California

Detection and interpretation of A, B, C, D from von der Haar et al., 1973, *Photogrammetria*, 29.

Satellite photograph:

Coast with Lagoons of Sonora, Mexico, around Guaymas and Gulf of California

Flight information data:

Mission: *NASA – APOLLO 7*
Date: *13. Oct. 1968*
Photograph: *colour*
Magazine: *5 frame no. 1632*
Camera: *Hasselblad 500 C*
70 mm photography
Lens: *Zeiss Planar f/2,8; 80 mm*
Film: *Kodak aerial Ektachrome*
Filter: *Wratten 2 A*
Altitude above mean sea level: *122 naut. miles*
Geographical position of the principal point:
latitude: *27 degrees 30 min. N*
longitude: *111 degrees 00 min. W*
Chart: *ONC – WAC code H – 22*
Local observation time: *12.36*
Sun azimuth: *180°*
Sun elevation: *55°*
sun glint (reflection) in the southern part of the photograph
Photograph viewing mode: *oblique exposure axis to 135°*
Cloud cover: *2 % and fog*
Scale of the printed picture: *variable*
Orientation of the printed photograph: *north is above.*
To regard this oblique photograph in direction of the camera axis turn the photograph 180° around.

NASA – APOLLO 7

Interaction of lagoonal waters with oceanic waters – complicated due to oceanic phenomena and processes change detection

Lagoons and Coast of Sonora, Mexico, around Guaymas and the middle part of the Gulf of California

NASA colour photograph, Apollo mission 7, BFC Photo no. 1632 MSC no. AS 7-5-1632; date 13. Oct. 1968; hand operated camera; 70 mm Hasselblad camera with ZEISS planar f : 2,8; 80 mm lens; oblique photograph; scale approximately 1 : 2 000 000, variable.
The northern third of the picture is virtually true to scale, the distorted southern portion rather smaller in scale. Exposure axis to the southeast. The photograph was taken about noon time (12.00 local time). So the azimuth of the sun is southerly. The sun's elevation is high – about 55 degrees, i. e. the photograph was taken with an exposure axis of 45 degrees against sun rays. This condition causes some reflection of the sun (sun glint) in the southern part of the photograph. But it is interesting that the phenomenon of sun glint occurs and is reported in most oblique photographs of orbital photography.
Orientation of the printed photograph: north is above. To regard this oblique photograph in direction of the camera axis turn the photograph 180° around.

Oceanographic phenomena considered in the study are circulatory systems, biological activity, turbidity, sea-surface state, slicks, upwelling, standing waves and nearshore bottom topography, and atmospheric feature cloud lines.

Description of the area

The picture encloses the Gulf of California around Guaymas. The southwestern corner shows the peninsula of Concepción with Concepción Bay on the peninsula of Lower California. In the northeast of the picture the coast of the state of Sonora with the Bay of Guaymas, the important port of Guaymas and the Rio Guaymas delta plain are identifiable. Near to the centre of the picture the triangular lagoon, Estero de Lobo, is identifiable along the coast. In the eastern part of the centre the irrigated agrarian country of Ciudad Obregón on the River Yaqui delta plain and the Alvaro Obregón dam are distinguishable on the northeastern margin. The small black wedge off the coast northwest of Guaymas is the small island of San Pedro Nolasco which is about 3 km long, situated about 20 km southwest of the coast of Sonora. Isla San Pedro Nolasco is a barren island of volcanic origin. It is 320 m high and has steep coasts. The illuminated coast is visible to the south (bright), the shadowed coast is visible to the north (dark). The island is identifiable due to its elevation, whereas on the surface of the sea low clouds seem to occur.
The harbour of Guaymas, the most important port in the Gulf of California, is an inlet on the eastern side of the peninsula of which Cabo Haro is the southern extremity.
Guaymas is situated at the mouth of the river Guaymas at the western end of the Bahia de Guaymas, a steep rocky bay surrounded by red volcanic cliffs. The bay which is between 5 m to 10 m deep, is enclosed by two islands. It is an important tourist centre for fishing, hunting and water sports. Guaymas is the centre of a shrimp-fishery, depending on the ecology of the coastal lagoons. In the middle of the Gulf there is the very important fishery for merlin.
The barrier island which closes up the eastern part of the Guaymas lagoon, is called Playa de Dolores, a narrow, sandy strip, visible on the photograph. The small flat island, Isla de Pajaros, in front of the entrance to Guaymas is discernable as a small grayish bar.

Coastal features

The coast north-west of Guaymas is mainly rocky and steep. The coast south-east of Guaymas is flat and bordered by lagoons with their barrier islands. These barrier islands, with sandy beaches (playas) and some small dunes, are visible on the satellite photograph as small white-pink lines and bands. Also, submerged sand bars, tidal channels and several tidal wash-over deltas are visible. Tobari lagoons lie on either side of Tobari Bay, between Punta Arboleda and Rio Viejo Mayo, the former mouth of Rio Mayo, 27 miles north-westward,

being separated from the sea by low sandy islands.
The coast between Tobari Bay and Punta Lobos, about 43 miles north-westward, is low, sandy and cut up by the entrances to the lagoons. Shoals extend from 2 km to 4 km off these entrances, and at about 6 km offshore, there are depths of from 11 m to 18 m. The islands change their positions with severe storms.

Punta Lobos is the north-western extremity of Isla Lobos, a low, sandy island, about 10 km long, which is separated from the mainland by Estero de la Luna: southward of Punta Lobos, there are depths of 11 m at about 1 km from the shore. Sedimentation in the narrows often inhibits the migration of fish and prawns between the lagoon and the sea.
The coast between Punta Lobos and the entrance to Rio Viejo Yaqui, about 45 km northward, is low and sandy, and backed by lagoons, none more than 2 metres in depth. Sandy shoals, with depths of less than 5 m over them, extend from 0.5 km to 2 km off this coast.
Strong currents, varying in direction, are found in this vicinity, and fogs are of frequent occurrence.

The river Yaqui is the most important in the area, but since the volume of fresh water at its mouth varies considerably on account of the irregularity of tributary waters and requirements for irrigation, the natural discharge varies considerably. Maps of the sediments of the gulf indicate that there are delta silty clays deposited as submarine fans in front of the river mouths. The construction of the Alvaro Obregón dam, which was completed in 1955, created a reservoir with a capacity of 3000 million m^3 (located 62 m above the level of the mouth of the river). In the coastal plain of the lower River Yaqui 200 000 hectares of land – supporting cotton and tomatoes – are irrigated from the river.
On the space photograph, the irrigated areas of Ciudad Obregón and Guaymas are identifiable.

The Gulf of California

In the water body of the Gulf, there is a rather complicated system of hydrographic processes, keeping water masses in motion depending on salinity, temperature, tides, upwelling waters and wind driven currents. Some of these effects can be observed on the satellite photograph.
Because interchange of lagoonal water masses is also involved, it seems necessary, to give a brief description of some of the natural features of the oceanography of the gulf. (For details see the extensive literature; in this paper only a few titles have been noted).
The Gulf of California is a marginal sea of the Pacific Ocean bordered by the Lower California peninsula in the west and in the east by the coasts of Sonora and Sinaloa of western Mexico. It is called Mar de Cortes, and at one time Mar Bermejo (Scarlet Sea), caused by plankton blooms, known as "red tide".
The Gulf trough is a tectonic formation stretching from 34° North for over 930 miles southwards.
From the mouth of the Colorado River it stretches 680 miles south-east with a width of about 37 miles and, at its conjunction with the Pacific, 130 miles. Area: 62 000 square miles, maximum depth: 3 200 m, in the Guaymas basin 1 800 m.
The tides: the Gulf of California, as a long narrow part of the Pacific Ocean, has particularly complicated tidal streams. While the maximum tidal range in the south is 1 m, in the north near the mouth of the Colorado it is on average over 8 m (max. 10 m). The tides are mostly of the mixed type, i. e. diurnal to semidiurnal with big differences from day to day.
In the central part of the Gulf, around the Guaymas area, the tidal ranges are: mean tidal range 1,5 m, spring tidal range 2,0 m, neap tidal range 0,5 m.

Salinity and currents

Apart from the tidal swell from the Pacific Ocean and the movements arising within the Gulf itself, the contrast between the practically rainless desert area in the north accompanied by a very high degree of evaporation, and the much wetter, tropical Pacific region in the south, plus freshwater influx, results in variable water circulation which can be measured from wide variations in salt content. The surface salt content in the northern Gulf is 35,7 ‰ and in the south 34,5 ‰. The surface water temperatures in the north (summer 30° C, winter 17° C) and in the central part around Guaymas (winter 17° C, summer 31° C) also show much greater seasonal differentiation than in the south. In the cape region it frequently happens that two water fronts form. Here the waters from the California current of the Pacific Ocean meet with the waters of the Gulf and of the southern oceanic region: the Gulf waters are of high temperature and high salinity whereas the waters of the California current are of low temperature and low salinity and the waters from the Cape Corrientes region in the south of high temperature and lower salinity. G. I. Roden, 1958, gives values for the exchange of waters between the Gulf and the open sea. He estimated that the average evaporation rate in the Gulf area is 5 000 m^3 per sec.

During the winter and spring when strong northerly winds blow in the Gulf, a considerable volume of water is thrust up to the surface. To balance this, ocean waters flow into the Gulf at depths between 500 and 800 m at rates of 350 m^3 per sec.

These various factors – tides and waters of different salinity and temperature – cause exchanges, interflows and cross-currents in the upper layers of the Gulf and the interaction of one with the other is highly complicated. In addition, wind-induced offshore currents occur.

Upwelling waters

On the Gulf coast of the peninsula there are a number of places where cold upwelling waters are prevalent (van Andel, 1964). With northwesterly winds, upwelling waters occur on the coast of Guaymas. This upwelling off the coast occurs in conjunction with the formation of fog banks which the author was able to observe several times in the vicinity of Santa Rosalía, Mulegé and Concepcion Peninsula (see southern part of the satellite photograph). Fog seems to cover sea surface around the rocky island San Pedro Nolasco.

Flying over the areas, it was possible on occasion to distinguish clearly the existence and movement of various water bodies, sometimes very sharply defined by tides and currents.

In the coastal zone, especially where river discharge and upwelling water ensures highly fertile environments, biological productivity of the ocean is at its greatest. Commercial fishing is important in this area.

Lagoonal water masses are involved in the intercharge of water circulation. (Compare photograph and interpretation).

In the Gulf Coast regions of California, there are a number of areas where fog occurs frequently. At Santa Rosalia, southwards at Mulegé and at Guaymas, there are often thick fog patches which can cause the air temperature to drop from 30° C to 15° C within a few hours. (Its occurrence is beneficial for the vegetation in these parts of the east coast). This fog is caused by cold upwelling waters which are the result of offshore winds and the meeting of various water masses in the tidal flux.

Interpretation of the satellite photograph

The satellite photograph from Apollo mission 7, taken 13. Oct. 1968, showing the Gulf of California in the Guaymas area, was, as one in a series, the object of a special study by S. P. von der Haar and R. O. Stone, 1973, published in "Photogrammetria". Some features of atmospheric and oceanic nature appear similar on this photograph.

The oceanographic features, as A, B, C, D, were detected and explained by these authors. Interpretation of the lagoonal waters has been made by Gierloff-Emden.

Since this satellite photograph presents some interesting oceanographic phenomena such as red tide, sea-surface state, internal waves and slicks, an interpretation of these is given below:

Red tide and horizontal dynamics of water masses

Sinuous tongues of discoloured water, south of Guaymas, are interpreted to be phytoplankton blooms, indicated by stream line features, such as slicks, on the surface of the sea. Of particular interest in connection with the vertical water exchange caused by upwelling near the coasts, is the occurrence of cold, highly nutritive water containing phytoplankton (particularly diatomeae and dinoflagellata) which gives a reddish colour to the surface waters. This is probably the reason why the Spaniards called the Gulf of California not only "Mar de Cortes" but also "Mar Bermejo" – the Scarlet Sea.

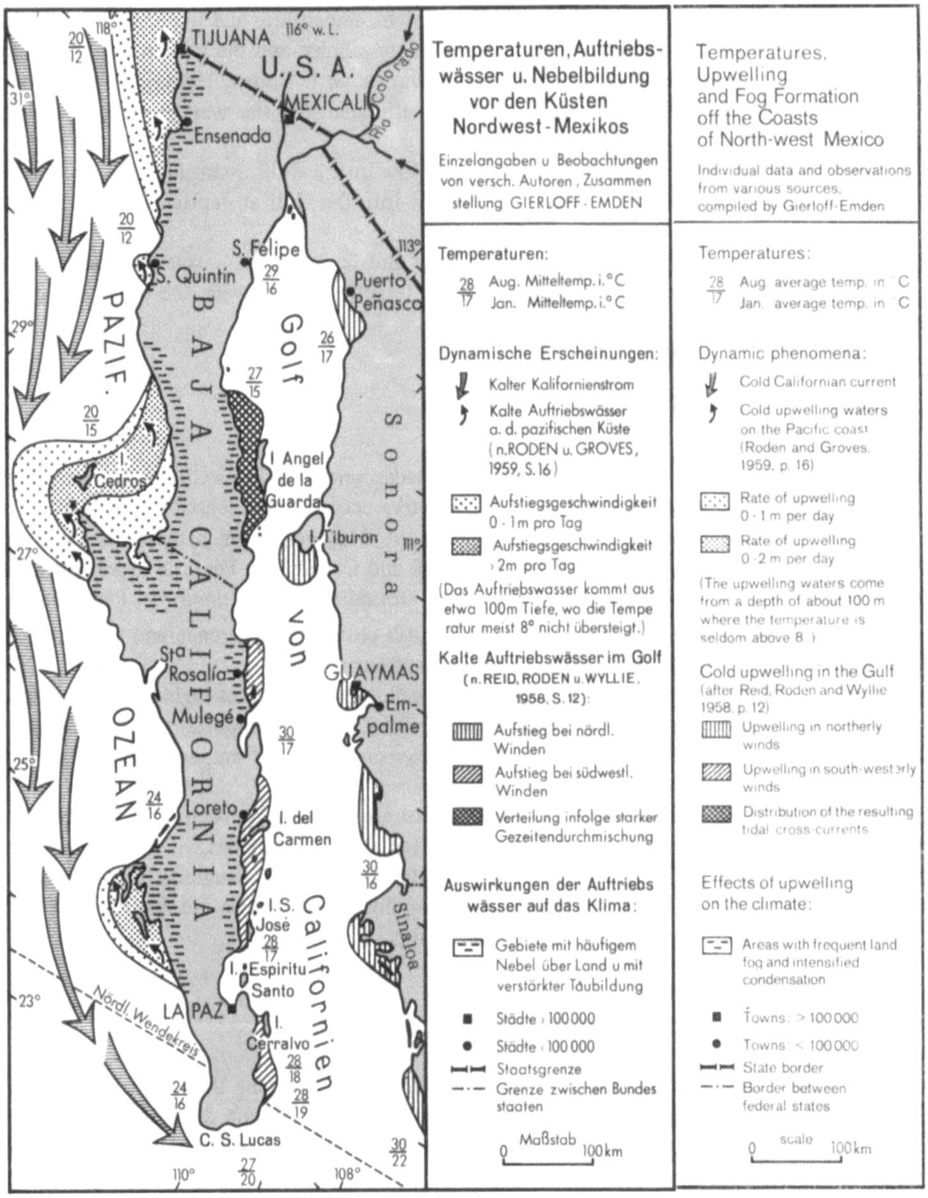
Temperaturen, Auftriebswässer u. Nebelbildung vor den Küsten Nordwest-Mexikos
Einzelangaben u Beobachtungen von versch. Autoren, Zusammenstellung GIERLOFF-EMDEN
Temperaturen:
28/17 Aug. Mitteltemp. i.°C
Jan. Mitteltemp. i.°C
Dynamische Erscheinungen:
Kalter Kalifornienstrom
Kalte Auftriebswässer a. d. pazifischen Küste (n. RODEN u. GROVES, 1959, S. 16)
Aufstiegsgeschwindigkeit 0 - 1 m pro Tag
Aufstiegsgeschwindigkeit › 2 m pro Tag
(Das Auftriebswasser kommt aus etwa 100 m Tiefe, wo die Temperatur meist 8° nicht übersteigt.)
Kalte Auftriebswässer im Golf (n. REID, RODEN u. WYLLIE, 1958, S. 12):
Aufstieg bei nördl. Winden
Aufstieg bei südwestl. Winden
Verteilung infolge starker Gezeitendurchmischung
Auswirkungen der Auftriebswässer auf das Klima:
Gebiete mit häufigem Nebel über Land u mit verstärkter Taubildung
Städte › 100 000
Städte ‹ 100 000
Staatsgrenze
Grenze zwischen Bundesstaaten
Maßstab 0 100 km
Temperatures, Upwelling and Fog Formation off the Coasts of North-west Mexico
Individual data and observations from various sources, compiled by Gierloff-Emden
Temperatures:
28/17 Aug. average temp. in °C
Jan. average temp. in °C
Dynamic phenomena:
Cold Californian current
Cold upwelling waters on the Pacific coast (Roden and Groves, 1959, p. 16)
Rate of upwelling 0 - 1 m per day
Rate of upwelling 0 - 2 m per day
(The upwelling waters come from a depth of about 100 m where the temperature is seldom above 8°)
Cold upwelling in the Gulf (after Reid, Roden and Wyllie 1958, p. 12)
Upwelling in northerly winds
Upwelling in south-westerly winds
Distribution of the resulting tidal cross-currents
Effects of upwelling on the climate:
Areas with frequent land fog and intensified condensation
Towns: › 100 000
Towns: ‹ 100 000
State border
Border between federal states
scale 0 100 km
TIJUANA
U.S.A.
MEXICALI
Rio Colorado
Ensenada
S. Felipe
S. Quintin
Puerto Peñasco
BAJA CALIFORNIA
PAZIF. OZEAN
Golf von Californien
Sonora
I. Cedros
I. Angel de la Guarda
I. Tiburon
Sta Rosalia
GUAYMAS
Empalme
Mulegé
Loreto
I. del Carmen
I. S. José
I. Espiritu Santo
LA PAZ
I. Cerralvo
Sinaloa
Nördl. Wendekreis
C. S. Lucas
116° w. L.
118°
113°
111°
110°
108°
31°
29°
27°
25°
23°

Interpretation of Red Tides, sea-surface state, standing waves
following S. P. von der Haar and R. O. Stone, 1973

Red Tides
"Plankton blooms or "red tides" in the upper Gulf of California, although viewed by numerous investigators, have received scant mention in the scientific literature. One major contribution is by Cannon [1966] who stated that the greatest plankton blooms occur in February and March at the time of extreme lunar tides. The water then becomes cloudy, brownish or brilliant green. Calvert [1966], however, cautioned that annual peaks and uniform distribution may be an oversimplification. More data on the type and quantity of nutrients necessary to trigger blooms and on their temporal, areal and vertical distribution are needed before orbital photography can be used to monitor accurately this important biological phenomenon.
No definite plankton blooms or other biogenic activity could be detected on the orbital photographs of the upper Gulf, but photographs of the central portion of the Gulf exhibit classic examples. A white scum line, indicative of biological activity, can be seen in the counter-clockwise eddy south of Isla Tiburon. Long sinuous tongues of discoloured water (photograph) in the sun-glint area south of Guaymas at 27.5° N are interpreted as phytoplankton blooms. Surface oil slicks in the same area are likely the result of related biological activity.
The variety of colours exhibited by blooms may pose a problem in identification from orbital photographs. Phytoplankton in the sea surface layers may not be seen on colour film because of a minor shift in the blue spectrum produced by organic constituents and infrared photography would not reveal deeper concentrations of plankton." Von der Haar and Stone, 1973.

Sea-surface state
(Interpretation of von der Haar). "The sea-surface state as evidenced by tone and texture changes, shows a variety of rough and calm patches, particularly in the sun-glint regions. McClain and Strong [1969] found that calm patches or strips in the diffuse glint region appear dark unless the specular point lies within the calm water, in which case it appears as an anomalous bright patch within the general glint area. These dark patches can be used to infer sea-state variations. Dark streak-like anomalies often correspond to calm waters (low sea-state) beneath high pressure ridges or when parallel to the coastline, the seaward limit of local sea breeze conditions." ... "Across the width of the Gulf is a sequence of a light-coloured area (rough seas), a dark-coloured irregular zone (smooth seas) and an adjoining light-coloured or rough area. The two rough surfaces represent water agitated by the sea breeze toward the Mexican mainland in one instance, and toward the Baja peninsula in the other."

Standing waves
"A standing wave is one which oscillates in a vertical fashion between two nodes or points and is a wave form which exhibits no progressive movement. Seven waves, with an approximate wavelength of 2 km, appear as parallel dark bands on the sea surface."
Remark: general description of such phenomenon see: E. P. McClain and A. E. Strong, 1969, On anomalous dark patches in satellite-viewed sun-glint areas: Mon. Weather Rev., vol. 97, p. 875–884.

Lagoonal waters near the coast of Sonora, a phenomenon of change detection
interpreted by Gierloff-Emden

On the satellite photograph, it can be observed that in the vicinity of the coastal lagoons – this can be very clearly seen around Isla Lobos – there is some discolouration of the water in the Gulf. Light blue tongues and plumes extend perpendicularly from the coast into the Gulf up to a distance of 25 km and more. This covers an area where the depth of the Gulf is greater than 200 m.
The question is, do these plumes represent debris laden, turbid water masses of lagoonal character, pouring out of the lagoon, or are some upwelling waters involved?
Because of the form of these tongues, as for instance near the inlet of Laguna Lobos, there seems to be a specific water mass, like a band. One can estimate, that at least the deep grey-blue bands have come from the lagoons. These bands, called tongues or plumes, due to their curved feature, end in two clockwise orientated eddies in the gulf waters (as described above).

Conclusion

These observations disprove the idea that lagoonal water masses, which are in interchange with the water masses in the Gulf, cover large areas and water bodies; in fact a very strong exchange can happen. This is important for the ecology of the lagoons in this specific case for the shrimp fishery. Current pattern along the Mexican coast can be mapped better with the help of orbital photography.

Interpretation of a satellite photograph can help to identify the catch or drainage area of lagoonal water masses in the Gulf and using this information, it would be easy to select stations for regional oceanographic studies or to plan oceanographic expeditions.

Pollution problems of the Sonora coast

In the Guaymas area, there are three sources of pollution:
1. sewage waters (aguas negras) from the larger towns;
2. agricultural waste, in this area from cotton and sugar production of the large irrigated delta region at Rio Yaqui and Rio Mayo in the form of cotton fibres and filaments; and
3. the residues of pesticides such as DDT and others, used in the extended irrigation areas as insecticides, washed downwards to the ocean by rain and rivers and blown by the winds.

Information is urgently needed, not only as to which of these pollutans are poured into the lagoons, but also how far these materials extend from the lagoons, with the outflowing waters into the ocean. (Compare satellite photograph).

Lit.: Cifuentes Lemus, J. L., R. R. Castro and A. Z. Menez, 1972, Panorama general de la contaminacion de las aguas en Mexico: in: FAO: M. Ruvio (ed.), Marine pollution and sea life, pp. 100–106. London: Fishing News (Books) Ltd.

Bibliography

Andel, T. van and G. G. Shor jr. (ed.), 1964, Marine geology of the Gulf of California, a symposium: *Amer. Assoc. Petroleum Geologists*, 408 p.

Byrne, J. V., 1957, The marine geology of the Gulf of California: thesis, Univ. South Calif., Los Angeles, 298 p. (unpublished).

Clarke, G. L., G. C. Ewing and C. J. Lorenzen, 1969, Remote measurement of ocean colour as an index of biological productivity: Proc. 6th Int. Symp. Remote Sensing of the Environment, Univ. Michigan, vol. 2, pp. 491–1001.

Ewing G. C., 1950, Relation between band slicks at the surface and internal waves in the sea: *Science*, 111, p. 91–94, New York.

Gierloff-Emden, H. G., 1964, Die Halbinsel Baja California: *Mitt. Geogr. Ges. in Hamburg*, vol. 55, 160 p.

Gierloff-Emden, H. G., 1970, Mexico, eine Landeskunde: de Gruyter Verlag, Berlin, 634 p.

Gilbert, J. Y. and W. E. Allen, 1943, The phytoplankton of the Gulf of California obtained by the E. W. Scripps cruise in 1939 and 1940: *Jour. Marine Research*, no. 5.

Haar von der, S. P. and R. O. Stone, 1973, Oceanographic analysis of orbital photographs of the upper Gulf of California: *Photogrammetria* 29, p. 45–61.

Munk, W. H., 1941, Internal waves in the Gulf of California: *Sears Found. Jour. Marine Sci.*, vol. 4, p. 81–91.

Phleger, F. and G. C. Ewing, 1962, Sedimentology and oceanography of coastal lagoons in Baja California, Mexico: *Geol. Soc. America*, Bull., vol. 73, p. 145–182.

Roden, G. I., 1958, Oceanographic and meteorological aspects of the Gulf of California: *Pacific Sci.*, vol. 12, no. 1, p. 21–45, Honolulu.

Roden, G. I., 1964, Oceanographic aspects of the Gulf of California: in Tj. van Andel and G. G. Shor jr. (eds.), Marine geology of the Gulf of California: *Amer. Assoc. Petroleum Geologists*, Mem. 3, p. 30–58.

Roden, G. I. and G. W. Groves, 1959, Recent oceanographic investigation in the Gulf of California: *Sears Found. Jour. Marine Research*, vol. 18, p. 10–35.

Römer, E., 1936, Örtlich-periodisch auftretende Kabbelungen an der mexikanischen und mittelamerikanischen Westküste: Ann. Hydrogr. mar. Meteorol., Spec. Pap. of Sept. no., p. 55.

Scripps, E. W., 1950, Cruise to the Gulf of California: *Mem. Geol. Soc. America*, no. 43.

Soules, S. D., 1970, Sunglitter viewed from space: *Deep Sea Res.*, vol. 17, p. 191–195.

Szekielda, K.–H., 1972, Ozeanische Strukturen in Satellitenbildern: *Umschau*, no. 3, p. 95–97.

West Coasts of Central America and United States Pilot, Hydrographic Department, Admiralty, London.

Maps

Chart ONC – WAC code H – 22

Nautical charts of the Hydrographic Department, Admiralty, nos. 2222, 2324. British.

Land, lagoon and ocean as a system of environmental problems (pollution)

Coast of Houston—Galveston, Texas, U.S.A. and Gulf of Mexico

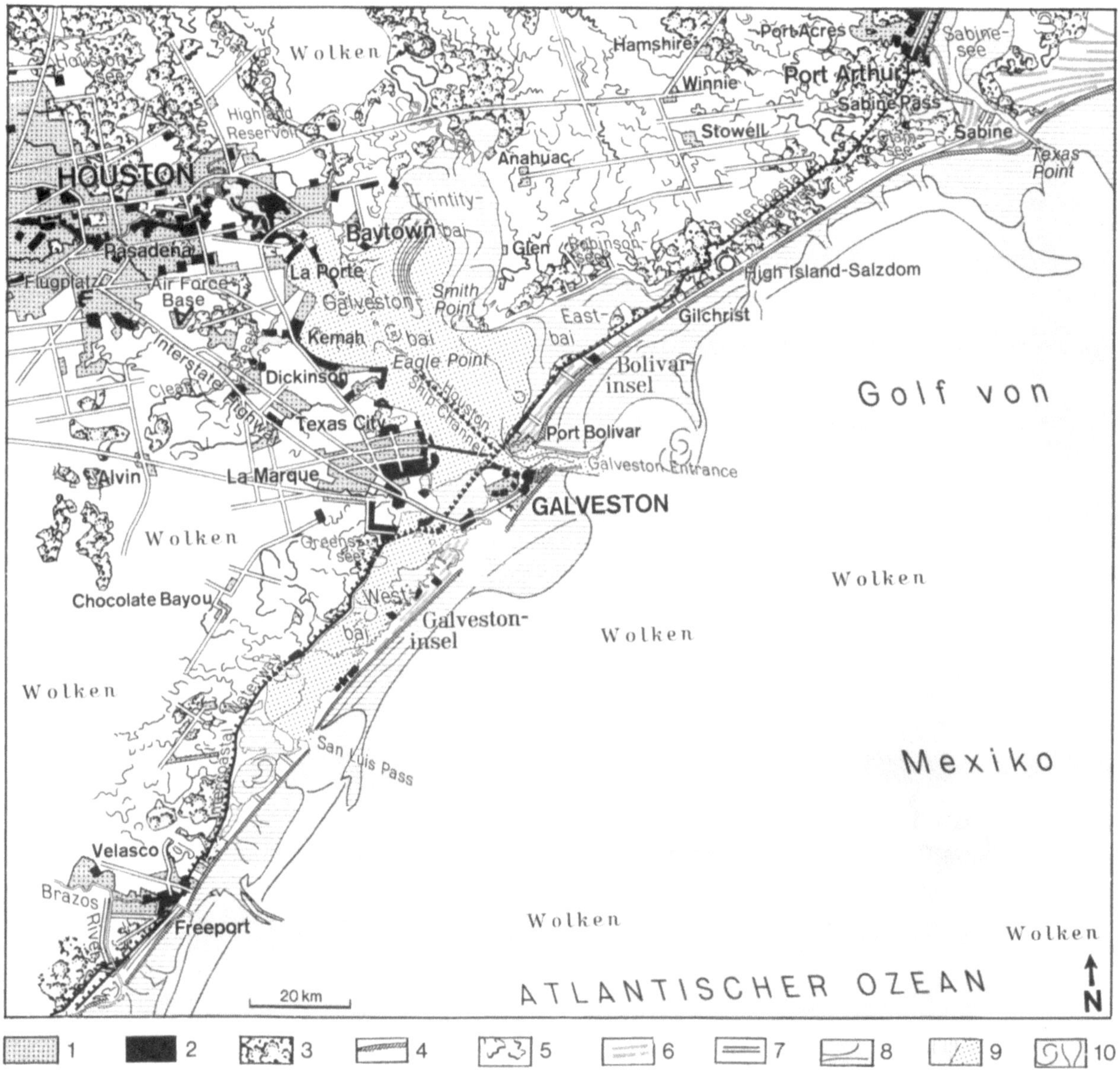

HOUSTON
Pasadena
Baytown
La Porte
Kemah
Dickinson
Texas City
La Marque
GALVESTON
Alvin
Chocolate Bayou
Velasco
Freeport
Brazos River
Port Arthur
Port Acres
Hamshire
Winnie
Stowell
Anahuac
Sabine Pass
Sabine
Sabine-see
Texas Point
High Island-Salzdom
Gilchrist
Port Bolivar
Bolivar-insel
Galveston Entrance
Galveston-insel
San Luis Pass
Trinity-bai
Galveston-bai
East-bai
West-bai
Smith Point
Eagle Point
Houston Ship Channel
Interstate Highway
Intercoastal Waterway
Flugplatz
Air Force Base
Wolken
Golf von Mexiko
ATLANTISCHER OZEAN
20 km
N
1
2
3
4
5
6
7
8
9
10

Coast and land of Texas, USA, around Houston–Galveston, Gulf of Mexico

Interpretation map, from NASA satellite colour photograph

APOLLO mission 9, no. C–3464

Interpreter H. G. Gierloff-Emden, 1975

Legend of topographic and hydrographic features classified in zones with selected discriminated objects (generalized: no differentiation of distinct water levels on this small-scale map)
Scale approximately 1 : 1 000 000

I above High Water level: terrestrial

mainland on the map as white area

a) coastal plain complex
(with natural and man-made environment)

1 settlement areas

2 industrial areas (partly altered in lagoonal environment)

7 highways, roads and streets

3 parkland, forests, partly swampy

5 bayous (partly to zone II and III)

b) coastal barrier complex

4) sandy beach on barrier islands

6 beach ridges

II between High and Low Water level: inter-tidal zone

no distinction of specific objects of this very zone

III below Low Water level: lagoonal and marine

(with natural and man-made environment)

8 shallow water areas in lagoons and in foreshore region with rip currents

9 lagoons, bays (estuaries)
partly with turbid waters in motion (indicates pollution)
with altered areas

11 Houston Ship Channel and Intracoastal Waterway as indicated with symbol on interpretation map

10 boundaries of water masses in motion
(stream lines, tongues, eddies)

Satellite photograph:

Coast and land of Texas, USA
around Houston–Galveston,
Gulf of Mexico

Flight information data:

Mission: *NASA – APOLLO 9*

Date: *9. March 1969*

Photograph: *colour*

Magazine: *C frame no. 3464*

Camera: *Hasselblad 500 C*
70 mm photography

Lens: *Zeiss Planar f/2,8; 80 mm*

Film: *Kodak Ektachrome*

Filter: *haze*

Altitude above mean sea level: *100 naut. miles*

Geographical position of the principal point:
latitude: *29 degrees 15 min. N*
longitude: *94 degrees 48 min. W*

Chart: *ONC – WAC code H – 24*

Local observation time: *13.20*

Sun azimuth: *190°*

Sun elevation: *50°*

Photograph viewing mode: *slightly oblique but near vertical*

Cloud cover: *18 % cirrus fibratus*

Scale of the printed picture: *1 : 1 000 000 (approximate)*

Colours of the printed picture (Munsell scale):
The printed picture is somewhat different from the NASA copy,

1. *land occurs more yellowish on printed picture*
   land occurs more greenish on NASA copy
2. *ocean occurs more dark on printed picture*
   ocean occurs more light blue on NASA copy
   and shows more transparency

NASA – APOLLO 9

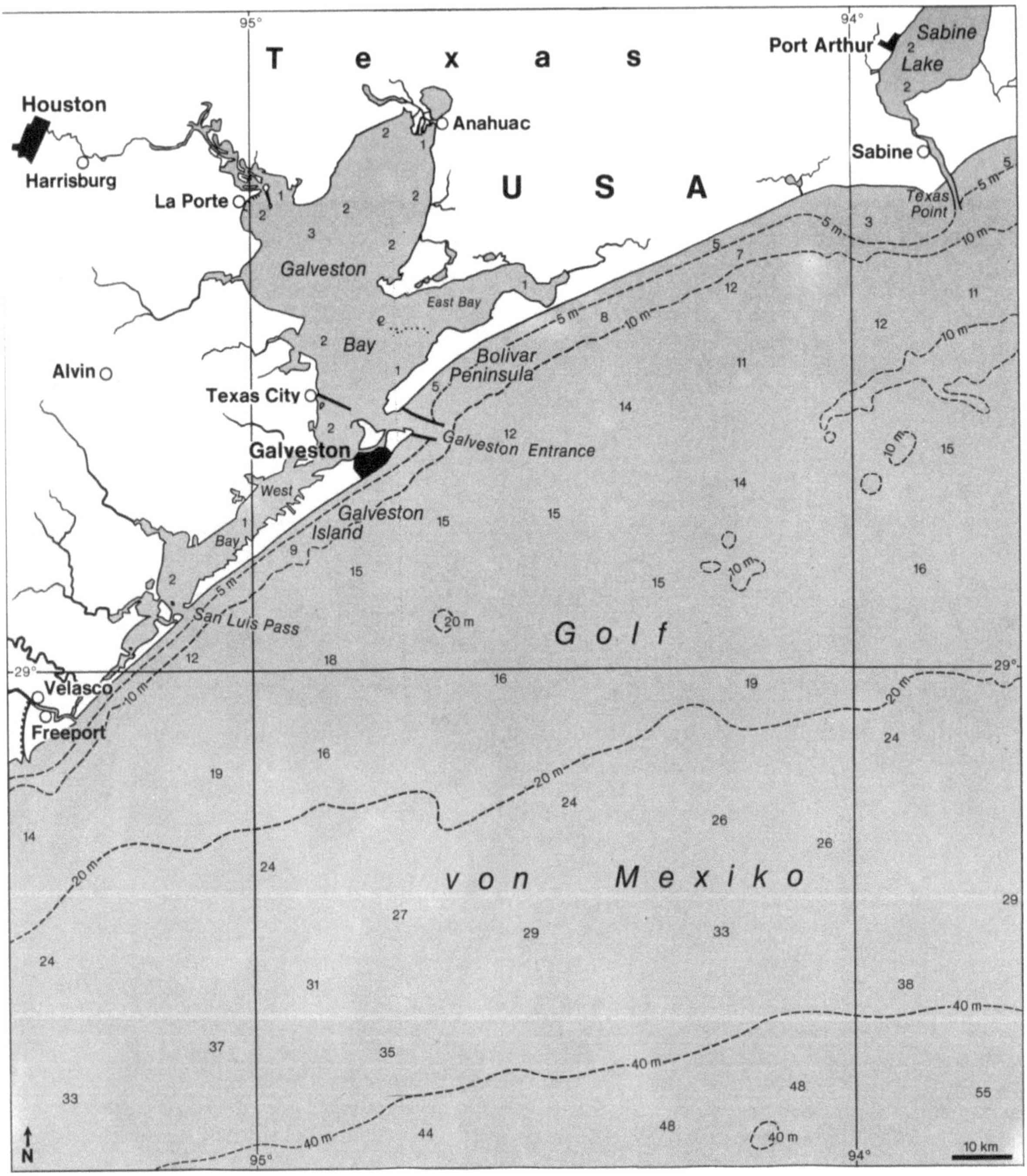

Coast and land of Texas, USA

Houston–Galveston, Gulf coast of Texas and Gulf of Mexico

Chart of the area of the satellite photograph
scale 1 : 1 000 000 (approximately),
with bathymetric topography, based on nautical charts

Land, lagoon and ocean as a system of environmental problems (pollution)

Coast of Houston–Galveston, Texas, U.S.A. and Gulf of Mexico

NASA colour photograph, Apollo mission 9, AS 9-3464; date 9. March 1969;
Houston–Galveston, Gulf of Mexico;
hand operated camera, camera Hasselblad 500 C, objective Zeiss planar f 2,8, 80 mm.
The scale of this photograph is approximately 1 : 1 000 000 (picture scale). The photograph is a slightly oblique one; there is a geographic distortion (the southern part of the picture is on a larger scale in comparison to the northern part). The photograph is approximately orientated to the north. Exposure axis to the north. The photograph was taken about 13.20 local time so that the azimuth of the sun is south-southwesterly. The sun's elevation is 50 degrees. Due to the direction of the exposure axis slightly with the sun light there is no glint effect and a remarkable water penetration is to be observed along the coast and partly in the Bay of Galveston.

The satellite photograph of the Houston–Galveston area of the Gulf Coast of Texas is one of the classic examples of space photography and its use in oceanography. It covers an area of very complex character, both as regards the land area and the water bodies. From the aspect of a geographer, this is a landscape where "man and environment", as Carl O. Sauer, world famous geographer from Berkeley, California, who died this year (1975) of an age of 85, once pointed out, becomes a system of natural and cultural relations. There are so many problems involved, that it is more important to identify some of these processes, and to demonstrate the complexity and the difficulty of interpreting the space photography, than to repeat some of the fine detail work, which can be done and is done by specialists on selected phenomena in this area.
The author visited the Gulf Coast for the first time in 1955 in company with Dr. Gordon Gunter (today in Ocean Springs, Mississippi), then again several times later; he had the opportunity to visit the late Prof. R. Russel and members of his staff from the Institute of Coastal Studies of Baton Rouge, La, including Prof. Walker and Prof. McIntire.
The most important impression obtained, was that this is an area, where the problems of our time, i. e. industrialisation, food gathering (fishing), recreation, pollution and natural processes, co-exist in a very complicated system.

Description of the area

Houston–Galveston – business district, industry on the Gulf Coast, lagoons and water masses.

The satellite photograph shows the Gulf Coast of Texas, an area approximately 180 km in length and width with land and sea; the printed photograph slightly less (approximately 170 km in length and width).
It embraces the flat Gulf Coast with its marshland, bays and lagoons and barrier islands; parkland, forest, swamp and townscape stand out clearly on the photo.
Galveston is located in the centre of the picture; in the lower left corner, Freeport; in the upper right, Sabine Lake and Port Arthur; to the left of the centre, Galveston Bay with the Houston Ship Channel leading to the city of Houston in the upper left corner of the picture.
Streets, industrial plants, docks and residential areas are visible in detail. Between Houston and Galveston, near Clear Lake, the area of NASA's Manned Spacecraft Centre is discernible. The photograph has an "X-ray"quality. The picture allows comparison of the contemporary landscape with maps which are inevitably dated and was used to update existing topographic maps.
The Gulf Coast plain occupies less than half of the Gulf coastal area of Texas and extends along the coastline in a belt approximately 100 km wide. This belt can again be divided into the actual coastal area – the bay and lagoon country – which reaches inland to the points and bays of the great lagoons, some 40 km from the Gulf, and into an inner belt which is known comprehensively as prairie land.
The coast is bordered by a band of narrow barrier islands with sandy beaches. These extend along 500 km of the 650 km length of the outer coast of Texas.
The Galveston Bay complex is the largest and one of the most important estuarine systems, both economically and ecologically, of all the bays on the Texas coast and covers 1 200 square km. The Galveston Bay estuarine system is composed of Upper and Lower Galveston Bays, Trinity Bay, East Bay and West Bay.

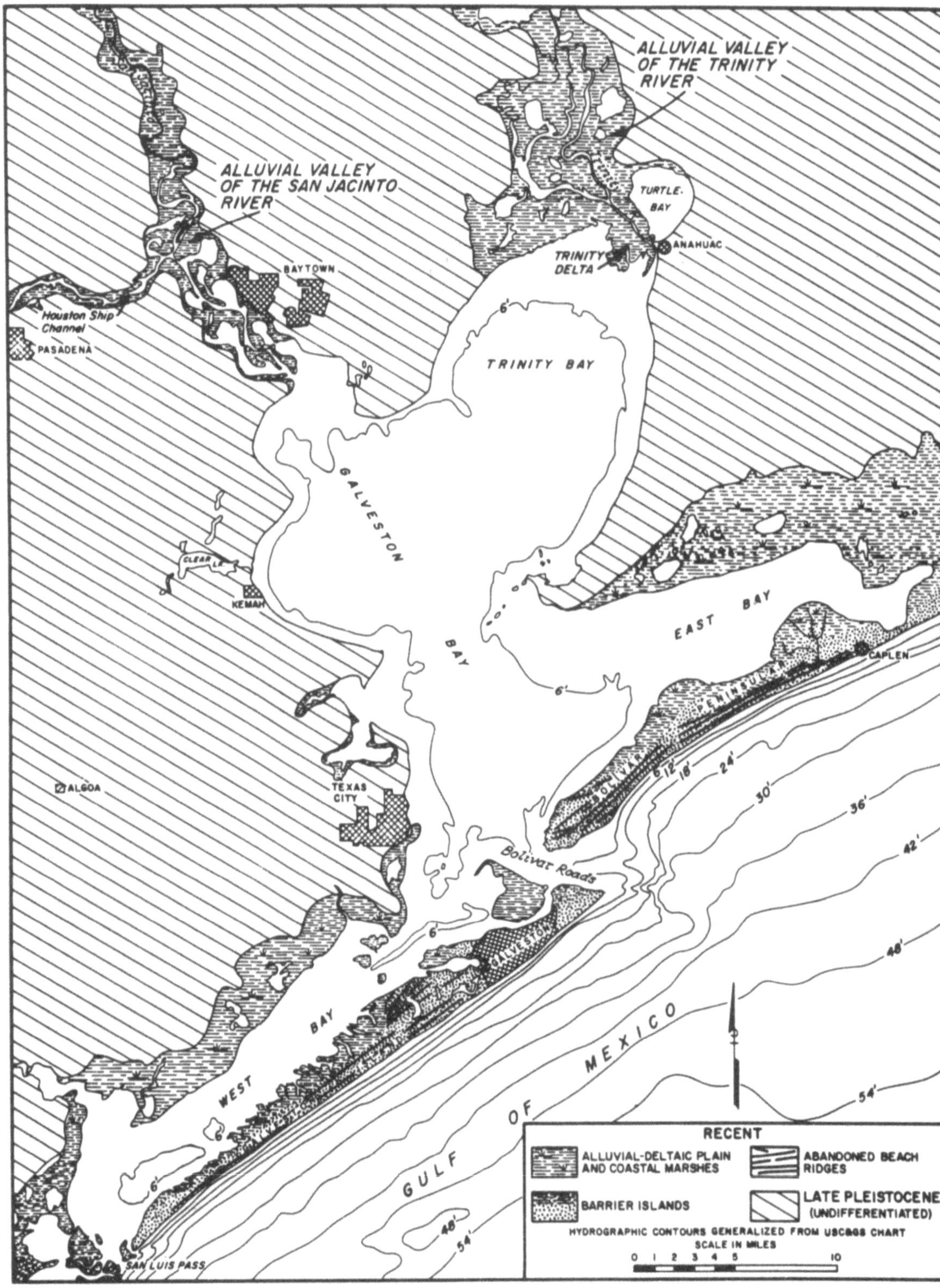

The Galveston Island-Bolivar Peninsula segment of the Texas shoreline

from R. J. Leblanc and W. D. Hodgson, 1959, Origin and development of the Texas shoreline: 2nd Coastal Geography Conference, Louisiana

Water depths in the bays range from 0.5 to 3 m, averaging about 2.2 m, giving a total volume of water at mean low tide of 3 000 million m^3. The bay is protected from the open Gulf by a long spit on the east, Bolivar Peninsula, and on the west by Galveston Island, a barrier.
For classification of features and sedimentary composition of the lagoon regions on the Gulf coast of Texas compare R. Leblanc and A. Bernard, 1954, Origin and development of the Texas shoreline: in 2nd Coastal Geogr. Conf., 1959, Baton Rouge.
Following R. Leblanc and W. D. Hodgson, one can classify:

1. the lagoonal type of bay; lies directly behind barrier islands and is elongated parallel to the coast,
2. the estuary type of bay; is elongated normal to the coast (bay is used as a general term, includes "estuary" and "lagoon").

In the Gulf of Mexico, the 10 m depth line is located about 3 nautical miles offshore and a depth of 100 m only at a distance of 150 km from the coast. One of the major characteristics of the bay system is the series of dredged channels and accompanying shallow spoil banks and islands, altered by man including the Intracoastal Waterway (4 m to 5 m depth), the Houston Ship Channel (11 m to 13 m depth), the Texas City Channel (11 m depth). The Houston Ship Channel, excavated to a depth of 11 m in the flat Galveston Bay, is only just discernible on the satellite photograph as a line of dark blue colour. Many old river courses, "bayous", are clearly discernible.
The natural resources such as mineral oil, natural gas, lime, sulphur and magnesium found in the Gulf Coast area of Texas have given rise to extensive industrialization. In the Gulf Coast area, numerous mineral oil fields with rich wells are to be found on salt domes.
"High Island was formed by a salt intrusion". It was one of the first oil-bearing salt domes to be discovered; (in 1901) production began in 1922. The top of the salt is about 360 m below sea level. High Island is a circular hill along the Texas coast 35 km east-north-east of Galveston (its height is only 6 m), situated close to the shoreline. High Island is visible on the satellite photograph as a small spot surrounded by roads.
Hurricanes – extremely fast moving cyclones – which hit the Gulf Coast of Texas from the south over the Caribbean Sea – have an important effect on climatic conditions and coastal features on the Gulf Coast. Since 1900, Galveston has been struck by severe hurricanes in 1909 and 1949, and, on 11. September 1961, by hurricane "Carla", which passed directly over the Houston-Galveston area. The coastal changes produced by more recent hurricanes are well illustrated by aerial photographs. F. P. Shepard, 1971, "Our changing coastlines".
On the satellite photograph, most of these features are of a smaller scale than the minimum visible, but some particulars are discernible.
Winds, storms, "northers" and hurricanes give rise to exceptional water levels and waves which break up the spits and form erosion channels and deltas (see photograph). In such cases the water on the coast (sea level) can reach a height of up to 5.5 m above the average high tide mark; when this happens, floods can reach up to 40 km inland and cause alluvial deposits. Several transverse furrows are evidently the sites of sluiceways across which the waters of hurricane tides have swept. Rainfall causes direct denudation of the land spits and isolated lagoons are washed out. This causes further silting up which, in its turn, results in changes in the whole ecological and biogenic scene.
An extensive storm warning system equipped with special radar stations has been constructed to protect the flat coastal land from the dangers of this elemental force. Satellite weather charts have also been used with success.

Fresh water and sediments from the rivers to the lagoon

The bay system receives its major source of fresh water input from the Trinity River, which flows into Trinity Bay, and from the San Jacinto River, which flows into the north-western part of Upper Galveston Bay. Additional fresh water is contributed via the Intracoastal Waterway and the numerous bayous which drain secondary lakes, bays and adjacent marshes.
Most of the Texas rivers enter the heads of estuaries where they build deltas; but two of the largest rivers, the Rio Grande and the Brazos, have filled their former estuaries into which they once drained. The river water is often laden with debris.

The Brazos River transports an average of 30 million tons of silt per year. The Colorado River transports an average of 9 million tons of silt per year.
This silt and the finer suspended material influence the colour of the water as a natural tracer. Large amounts of suspended material are mixed with the water by dredging operations.
Note the colour variation apparent in the water, on the photo especially near the mouths of the major rivers and bays i.e. tidal inlets.

Urban development and dynamic landscape modification

The urban development has been particularly rapid and has changed the whole landscape. In some areas there is such a concentration of industries that one can indeed speak of an industrial landscape; this is the case, for example, in the area of the Houston Ship Channel, around Texas City, Corpus Christi and Port Arthur. Houston, with over 1,5 million inhabitants, has become the largest city in the South and the banking and business centre of the industrial Gulf Coast area of Texas and one of the biggest harbours of the USA. The well-developed road network, particularly the "highways", are clearly distinguishable in the satellite photograph. The "city centre", which is recognizable in the photograph by its surrounding ring of highways, is not very large, but the settled area of the city is quite extensive. Enormous amounts of domestic sewage are produced in this area, causing pollution in the river and the bay.
The plants of the petrochemical industry and mineral oil refining, particularly along the Ship Channel, steel, agricultural and metal industry, as well as machine engineering industry, are recognizable in the satellite photograph as grayish-white rectangles. In the environs of Galveston and the beach south-west of the town, and also over the waters of the Gulf, there is often a greenish-grey haze caused by industrial smog.
Port Arthur, a harbour and oil-industry town, is located at the crossing point of important shipping lanes: the canals to Beaumont and Orange run through here as well as the Gulf Coast Intracoastal Waterway.

Towns, settlement and industry and the ecology in the Galveston Bay area

Rapid and extensive industrial development, as experienced in this area, naturally involves great changes in settlement of the land and ecology of lagoons and bays in hydrological and biological components of the natural system of River-Bay-Ocean with its water mass interchange.
The whole system is now a function of man and environment, of a former system of natural conditions and a response on impact of man.
People and industry are concentrated into metropolitan areas centered around harbours but separated from each other by relatively undeveloped marsh or beach areas. Accessible beaches are rapidly turning into thriving and expanding resort communities of high seasonal population density and little industry. The coastal zone also serves as a recreational site for boating, fishing, hunting and trapping.
In 1900 Texas had a population of merely 3 million, in 1920 4,5 million. By 1954 it had grown to 8,5 million. The population of the forty counties on the Gulf Coast of Texas almost doubled from 1,365,174 in 1930 to 2,548,000 in 1954.
Sixty-six percent of Louisiana's population is near the marsh, and an estimated 10 000 000 man-days of sport activity annually take place in this environment.

"Estuaries along the United States coastline are being altered extensively by Federal, State and private institutions. More than 81,000 ha of shallow coastal bays (excluding marshes) in the Gulf of Mexico and South Atlantic areas have been altered over the past 20 years (Chapman, 1968).
In Texas deepening of about 700 miles of Federal navigation channels has altered 5,265 ha of bay bottom and destroyed 2,830 ha of brackish marsh and the dredged spoil has filled 2,025 ha of shallow bay and covered 9,315 ha of brackish marsh. Presently, large areas of shallow bays and marshes are being developed for waterfront housing sites. This involves dredging, bulkheading and filling. With expanding human populations and increased leisure time it is likely that the demand for these areas will increase.
When areas of shallow marsh and bay are deepened or filled with spoil, major changes in the bayshore environment include: (1) reduction in acreage of shore zone and marsh vegetation, (2) changes in marsh drainage patterns and nutrient inputs and (3) changes in water depth and substrates. The effects of these changes on estuarine organisms are poorly understood." W. L. Trent et. al., 1972.
"For the past fifty years or so the Texas bays have been the source of material known as "mudshell". It is dredged up from layers in the bottom of the bays, where it is covered with mud. The dredges frequently operate down to a depth of twenty-five feet below the surface of the bottom and it is known that some of

the deposits extend several feet deeper. The material is almost pure oyster shell, with very little admixture of other species. It is commercially valuable as a source of almost pure calcium carbonate and as a road material.
The reports on time when mudshell dredging began, the maps of old reefs now dredged up, and production records, especially in earlier years, are not very exact or are missing entirely. Nevertheless, millions of cubic yards of this material have been dredged from Texas bays in the last thirty years." G. Gunter, 1952, p. 134.

Pollution, ecology and detection of water masses

Industrial waste carried down by the rivers pollutes the lagoons and the Gulf (compare discolouration of water).
Much work has been done on the pollution problem of this area, but it is not known what influence pollution has on specific regions of the bays and on specific cycles of aquatic life.
The Texas Game and Fish Commission has been trying to make an inventory of these effluents. Because of its large area and variety of habitats, the bay provides nursery grounds for over 80 % of the poundage taken as fishery products in the Gulf of Mexico adjacent to the Texas coast.

"The Houston urban area is the fastest growing one in the United States. This may account for the fact that thirty-eight percent of the sports fishing in Texas is done in the Galveston Bay area. The state total sports catch of four common fishes, redfish, speckled trout, drum and flounder, last year was thirty-five million pounds. Of this amount 36,5 percent came from the Galveston Bay area. Thirty-eight percent of the fishing and 36,5 percent of the fish does not indicate any great decline in the fish population in the Galveston area, compared to the remainder of the Texas coast, most of which is totally lacking in either industrial or sewage pollution." G. Gunter, 1959.

Models of ecological systems

Studies of ecological responses of organisms to pollutants in aquatic systems have followed several different approaches, with most having the common objective of describing qualitative changes which have occurred in the biological community due to pollution. Recently, models on functional classification of ecological systems have been worked out.
As part of the Texas Water Quality Board's extensive Galveston Bay study, ecological investigations were accomplished in Galveston Bay during 1969 to assess the relationships between ecological response as manifest in species diversity, and the quantity and quality of waste water inputs into that estuarine system.
The primary objective of the study is to provide ecological cause and effect relationships for future modelling of waste inputs.

Field studies on pollution

"The relationships between species diversity of phytoplankton, zooplankton, nekton and benthos samples and the water quality of Galveston Bay, Texas, were quantitatively compared. Two water quality parameters (i. e., percent waste water or the toxicity of the water at a given sampling station) were found to be inversely correlated with species diversity.
A computerized dispersion model was used to calculate percent waste water. Toxicity was measured by subjecting a blue-green alga (Coccochloris elebans) to water samples and by utilizing growth depression as the toxicity criterion). Waters from the Houston Ship Channel and the Trinity River were found to be significantly toxic and the dispersion model was again used to determine the distribution of toxicity to each bay area. Evidence that toxicity was dispersed with input waters was experimentally verified. Those areas receiving the greatest amounts of toxic effluent exhibited the lowest mean annual diversities.
Predictive equations were generated, indicating that diversity at any point in the bay can be computed from Houston Ship Channel diversity and the dispersion model. Computations were also accomplished whereby dilution and / or treatment of waste inflows could be estimated to achieve a desired and / or acceptable diversity level in the bay." I. H. Blifford, jr. and D. A. Gillete, 1971, Species Diversity and Water Quality in Galveston Bay: Texas. Water, Air, and Soil Pollution 1, 1971, 89–105 by D. Reidel Publ. Company, Dordrecht-Holland.

Isopleths of the mean annual distribution of relative toxicity values in the Galveston Bay estuarine system, based on measurement of field work, indicate that the toxicity values are twice as high near the Galveston area than in the innermost part of the Bay near Trinity River estuary.

Exchange of water masses between lagoons and the ocean

The water bodies of the Texas lagoons contain the well known and important oyster-fishery, as well as the greatest shrimp fishery in the world today. Important in the lives of these shrimp is the cycle of their migration, as post larvae into the coastal estuaries, and as adults out of the estuaries. The adults produce their eggs in the coastal water areas of the ocean, i. e. in the Gulf of Mexico.
A detailed knowledge of the coastal currents will help to determine the migration paths of the shrimp to and from the estuaries and assist in predicting their abundance and distribution. The distribution of suspended sediments in coastal waters and the variations in the resulting turbidity of the waters are of great use in determining coastal water movements. The tidal range is only about 0,6 m but in conjunction with the fresh river waters flowing into it, is nevertheless effective concerning movement of water masses, sediments, submarine geomorphology and aquatic life.
In the northern region of the Gulf of Mexico, near the Texas coast, the prevailing direction of the surface currents throughout the year is from east to west (compare data maps from the US Naval Oceanographic Office). The velocity of flood- and ebb-streams varies locally; velocities of such currents have been observed up to 2 and 3 knots near the tidal inlets.

The use of satellite photographs to environmental problems

Tides, winds, fresh water from the rivers, with sediments in suspension and polluted waste from industry, generate together a rather complicated hydrological model of water masses in motion. Different transparency and water depth produce a picture of a large variety of coloured water. Nevertheless, some phenomena and processes can be observed on the satellite photograph and interpreted as processes of water masses in motion. Eddies of suspended material in the water can be detected in Trinity Bay and at the mouth of Galveston Bay. Plumes of turbid waters from the tidal inlets being deflected by coastal currents are visible.
A strip of greenish coloured water, a mixture of river and lagoon water, stretches about 25 km out to sea. This is of great importance to both, the fishery and pollution activities in the Gulf, and the satellite photograph can help to map the limits of the water masses simultaneously.
In shallow water, sand banks are discernable. The regular form of the lagoon outlet waterways can be seen (as depicted in models by hydraulic engineers for experiments): these are tunnel-shaped with bars and wing-shaped banks of sediment on the sides.
This single satellite photograph shows different phenomena and processes of different scales in unit time and metric scale; it suggests the complexity of the system of this landscape and man-made environment. Systematic analysis calls for space photography and imagery with different sensor systems and a selected repetition rate.
Concerning the mapping of small scale physical and cultural features which cannot be detected with this type of orbital remote sensing in this very area, it is neccessary to use high altitude flight aerial photography with orbital photography as an additional and useful tool.

Bibliography

The Texas coast has been studied for years by geologists of many oil companies. Notable among these are R. J. Leblanc, H. A. Bernard, and especially Harold Fisk, Richard Russel, Francis P. Shepard.
Gordon Gunter has published many papers on the fishing and ecology of the bays and lagoons of the Gulf Coast. Numerous scientific studies have been made. Only a very small selection is given below:

Arbingast, S. A. and J. R. Bradley, 1969, Socio-economic study Galveston Bay area: Rep. to *Texas Water Quality Board*, Austin.

Arnold, J. A. and A. H. Thompson, 1969, Apparent coastal water discolorations from space, oceanic and atmospheric: in P. C. Badgley (ed.), Oceans from space.

Bernard, H. A., R. J. Leblanc and C. F. Major jr., 1962, Recent and pleistocene geology of Southeast Texas: in *Geol. Soc. America*, excursion guidebook, Geol. of Gulf Coast and Central Texas, 1962 Ann. Meeting, p. 175–224.

Blifford, I. H. jr. and D. A. Gillete, 1971, Species diversity and water quality in Galveston Bay, Texas: in Water, air and soil pollution 1, p. 89–105, D. Reidel Publ. Company, Dordrecht, Holland.

Collier, A. and J. W. Hedgpeth, 1950, An introduction to the hydrography of tidal waters of Texas: *Publ. Inst. Marine Sci.*, Univ. of Texas, vol. 1, no. 2, p. 121–194.

Copeland, B. J. and E. G. Fruh, 1970, Ecological studies of Galveston Bay 1969: Rept. to *Texas Water Quality Board*, Austin.

Copeland, B. J. and T. J. Bechtel, 1971, Species diversity and water quality in Galveston Bay, Texas: in Water, air and soil pollution 1, D. Reidel Publ. Company, Dordrecht, Holland.

Fisk, H. N., 1959, Padre Island and the Lugana Madre flats, coastal South Texas: *Nat. Acad. Sci.*, Nat. Research Council, 2nd Coastal Geogr. Conf., p. 103–151.

Geyer, R. A., 1951, Bibliography of oceanography, marine biology, geology, geophysics and meteorology of the Gulf of Mexico: *Texas Jour. Sci.*, vol. 2, no. 1, p. 44–92.

Gunter, G., 1959, Pollution problems along the Gulf Coast: Trans. 2nd Seminar on biological Problems: in Water pollution, U. S. Public Health Service. Remark: Many other papers on this subject by G. Gunter.

Gunter, G., 1952, Historical changes in the Mississippi River and the adjacent marine environment. *Publ. Inst. of Marine Sc.*, issue vol. I no. 21.

Marmer, H. A., 1954, Tides and sea level in the Gulf of Mexico: in Gulf of Mexico, its origin, waters and marine life: U. S. Dept. Interior, *Fishery Bull.* 89 of *Fish and Wildlife Service*, vol. 55, p. 101–118.

Mash, F. D. and W. H. Espey, 1967, Shell dredging – a factor in sedimentation in Galveston Bay: Tech. Rept. to *Texas Parks and Wildlife Dept.*, Austin.

Norris, R. M., 1953, Buried oyster reefs in some Texas bays: *Jour. Paleont.*, vol. 27, no. 4, p. 569–576.

Price, W. A., 1947, Equilibrium of form and forces in tidal basins of coast of Texas and Louisiana: *Bull. Amer. Assoc. Petroleum Geologists*, vol. 31, no. 9, p. 1619–1663.

Report to the Galveston Bay study project, *Texas Water Quality Board*: in Ecological studies of Galveston Bay, 1969, supported by funds provided by the Texas Water Quality Board, Contract IAC (68–69)–408. The work was done at the University of Texas Marine Sci. Inst. at Port Aransas, Contrib. no. 2: Pamlico Marine Laboratory, N. C. State University.

Reynolds, T. D. and W. W. Eckenfelder, 1970, Reaction rates of Houston Ship Channel waters: Rept. to *Texas Water Quality Board*, Austin.

Russel, R. J. (ed.), 1959, Second Coastal Geography Conference, held on April 6–9, 1959, at the Coastal Studies Institute, Louisiana State University, sponsored by the Geography Branch, Office of Naval Research and the Nat. Academy of Sciences-Nat. Research Council, Committee on Geography, Advis. to the Office of Naval Research, Washington D. C.: with many papers on the Gulf Coast and numerous literature.

Shepard, F. P., 1956, Hurricanes affecting the coast of Texas from Galveston to the Rio Grande: *Beach Erosion Board*, Corps Engrs., U. S. Army, Tech. Mem. 78.

Shepard, F. P., 1958, Sedimentology and quaternary geomorphology of South Texas: *Gulf Coast Assn. Geol. Soc. Trans.*, vol. 8, p. 41–75.

Shepard, F. P. and D. G. Moore, 1955, Central Texas coast sedimentation, characteristics of sedimentary environment, recent history and diagenesis: *Amer. Assoc. Petroleum Geologists Bull.*, vol. 39, pt. 2, no. 8, p. 1463–1493.

Shepard, F. P. and D. G. Moore, 1960, Bays of Central Texas coast: in Recent sediments, northwest Gulf of Mexico, 1951–1958: *Amer. Assoc. Petroleum Geologists*, p. 117–152.

Shepard, F. P. and H. R. Wanless, 1971, Our changing coastlines, Miami; with numerous literature.

Shih-ang Hsu, 1970, Coastal air-circulation system: observations and empirical model: *Monthly Weather Review*, vol. 98, no. 7, July 1970, Coastal Studies Institute, Louisiana State University, Baton Rouge, La.

Stevens, C. S., 1951, The silt load of Texas streams: *Texas Jour. Sci.* vol. 3, no. 2, p. 162–173.

Trent, W. L., E. J. Pullen and D. Moore, 1972, Waterfront housing developments, their effect on the ecology of a Texas estuarine area: in M. Ruivo (ed.), Marine pollution and sea life: FAO.

Maps

Chart ONC – WAC code H – 24

Charts of the U. S. Naval Oceanographic Office.

U. S. National Ocean Survey Chart no. 1117, Galveston to Rio Grande, in scale 1 : 460 732.

German nautical chart, Deutsche Seekarte no. 912, DHI.

German nautical chart, Deutsche Seekarte no. 529, DHI, in scale 1 : 1 000 000 (Golf v. Mexico, NW-Teil).

Multispectral photography as a tool for interpretation of coastal phenomen
Differentiation due to different spectral reflectance of objects
(change detection and coastal boundary surveys)

Coast of Texas from Galveston Bay to Matagorda Bay, Gulf of Mexico

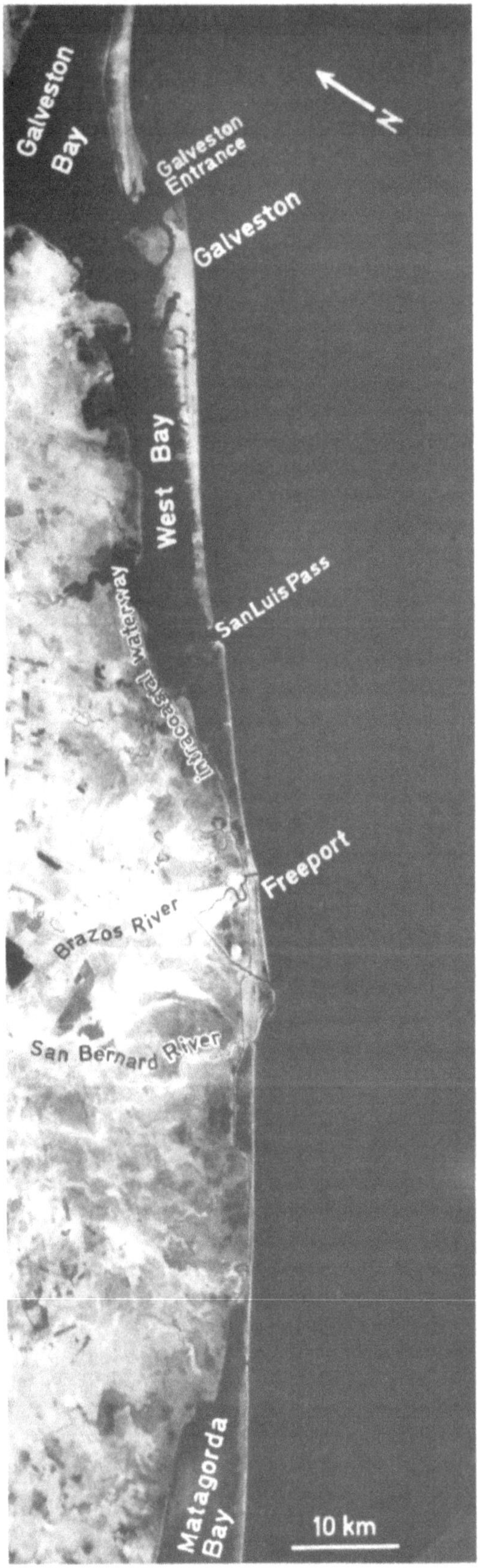

AS 9-26-A-3728 C
black and white film infrared

AS 9-26-A-3728 B
black and white film panchromatic

Satellite photograph:

Coast of Texas from Galveston to Matagorda Bay, Gulf of Mexico

Flight information data:

Mission: *NASA – APOLLO 9*

Date: *8. March 1969*

Photographs: *S 0–65 Experiment Multispectral Photographs*
AS 9

Magazine: *26 A* *frame no. 3728 C and 3728 B*

Camera: *Hasselblad 500 C*
70 mm photography

Lens: *Zeiss Planar f/2,8; 80 mm*

Film: *C: infrared black and white*
B: panchromatic black and white

Filter: *C with Wratten 89 b, B with Wratten 58 b*

Altitude above mean sea level: *100 naut. miles*

Geographical position of the principal point:
latitude: *29 degrees 04 min. N*
longitude: *95 degrees 24 min. W*

Chart: *ONC – WAC code H – 24*

Local observation time: *13.45*

Sun azimuth: *200°*

Sun elevation: *54°*

Photograph viewing mode: *near vertical*

Cloud cover: *2 % cirrus*

Scale of the printed pictures: *1 : 600 000*
(small sections of satellite photographs)

Satellite photograph:

Coast of Texas from Galveston to Matagorda Bay, Gulf of Mexico

Flight information data:

Mission: *NASA – APOLLO 9*

Date: *8. March 1969*

Photograph: *S 0–65 Experiment Multispectral Photographs*
AS 9 Infrared colour

Magazine: *26 A* *frame no. 3728 A*

Camera: *Hasselblad 500 C*
70 mm photography

Lens: *Zeiss Planar f/2,8; 80 mm*

Film: *infrared colour, mean wavelength of sensivity: green, red, infrared*

Filter: *Wratten 15*

Altitude above mean sea level: *100 naut. miles*

Geographical position of the principal point:
latitude: *29 degrees 04 min. N*
longitude: *95 degrees 24 min. W*

Chart: *ONC – WAC code H – 24*

Local observation time: *13.45*

Sun azimuth: *200°*

Sun elevation: *54°*

Photograph viewing mode: *near vertical*

Cloud cover: *2 % cirrus*

Scale of the printed picture: *1 : 600 000*
(section of satellite photograph)

NASA – APOLLO 9 AS 9-26-A-3728 A colour infrared

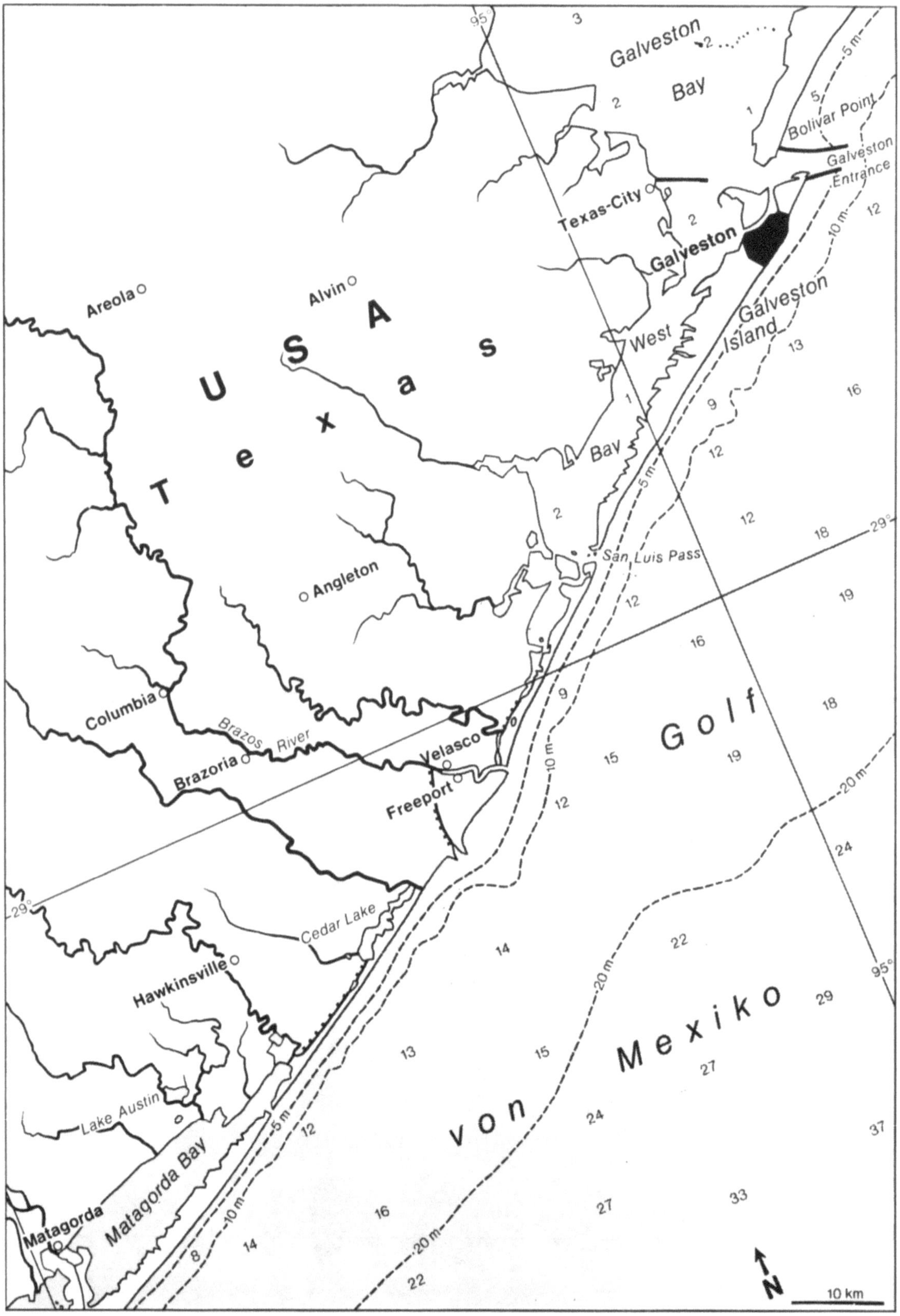

Coast of Texas from Galveston to Matagorda Bay

Chart of the area of the satellite photograph
scale 1 : 600 000,
with bathymetric topography, depths in m

Multispectral photography as a tool for interpretation of coastal phenomena. Differentiation due to different spectral reflectance of objects (change detection and coastal boundary surveys)

Coast of Texas from Galveston Bay to Matagorda Bay, Gulf of Mexico

NASA multispectral photography, Apollo mission 9, SO-65 experiment, camera axis near vertical, scale of the picture approximately 1 : 600 000. Photography with 4 lens Hasselblad camera with Zeiss-Planar objectives.

4 films are exposed at the same time from the combined camera:

A colour infrared film (Kodak-Ektachrome-Infrared-Aero), sensitive for the whole spectrum from 0,5 μm to the near-infrared 0,9 μm, with orange filter;

B black and white film, panchromatic with red filter;

D black and white film, infrared with red filter;

D black and white film, panchromatic with green filter (not reproduced here).

Photograph A is shown in its frame (except small strips at the margins).
Photograph C and B are shown as coastal strips.
In **photograph C**, being black and white infrared, all water surfaces appear black due to total absorption; the land provides a sharp contrast. As a result, the land-water boundary shows up sharply and clearly.
In **photograph B**, being panchromatic, penetration takes place through the water bodies; debris laden water masses show up grey to grayish white. These appear sharper than on the combined infrared colour film A, due to better differentiation between the green and blue bands. It is also an effect of the filter.
In **photograph A**, being colour infrared film, there are numerous gradations of colour density, due to full penetration to the bottom in very shallow water covered areas and the debris laden water masses.

Coastal features and coastal boundary surveys

There are differences in spectral reflectance of coastal areas so it is necessary to find out the optimum remote sensing media for differentiation and comparison. Changes in the same area as depicted by different remote sensing media, reveal different phenomenon. Interpretation using the colour photograph A only involves all the problems of photography through water bodies. (Compare example 1, Gulf of Venezuela, and example 2, Gulf of Guacanayabo in this book).
For instance: The small sand bar in front of the mouth of San Bernard River comes out grey-white, the same as do some debris laden water masses around it. The information is therefore ambiguous. Moreover, the bar extends as a submarine topographical object.
The infrared photograph C however clearly demonstrates that this white spot must be above the sea level (at the time when the photo was taken), i. e. dry, not covered with water. The debris laden water masses around are not visible on this photo, because with this film there is no penetration through the water.
As another example, compare the northern part of Matagorda Bay with the Intracoastal Water Way:
On the coastline, there is a (dry) beach and a submarine littoral (below water) beach. On colour photograph A and the panchromatic black and white B, both objects come out together as a white line. Because of the transparency of the water along the coast, it is not possible to distinguish between the dry beach and the submarine littoral beach, which is covered with a thin water mass.
On the black and white infrared photo (C), a smaller white beach line occurs representing the dry beach only.
Conclusion: With use of the black and white infrared film, one can decide with certainty between water covered and dry areas. Due to this selective process, in comparison with the colour photo, additional information is provided for mapping and detection without fear of error.
Remark: The coastal landscape has many geomorphological features and patterns. Many of these are objects of some 10 to 100 m in size and therefore not visible on the photographs (as for instance small beach ridges and small bars); they cannot be discussed here. The famous book from F. P. Shepard and

R. Wanless, 1971, "Our changing coastline" contains a series of large scale aerial photographs, some being of this region, for instance the mouth of the Brazos River (pp. 233). On these aerial photographs, the small scale objects are visible.

Oceanic features and change detection of suspended sediment discharge into shallow waters of the Gulf

Of particular interest is the detail shown in the colouration of the waters nearshore and offshore on the continental shelf. (The area includes depths inside the 40 m line only, mostly inside the 20 m line). (Compare colour infrared photograph A).

The most obvious features are the regions of outflow at the mouths of the bays. The debris laden turbid water and silt discharge and the shallow water regions of the littoral show up quite clearly as light grey and white bands; more sinuous tongues can be seen at the mouths of the bays, more straight bands and furrows at the littoral, from the beach to the ocean, having a general direction towards the south-south-west (SSW), at an angle of 30 degrees to the coast. This together indicates a south-westward drift of water masses with sediment particles in suspension. Because there is some penetration through the water, this is not only a surface flow. However, no rip currents can be observed. Northerly winds occur very frequently at this time of the year. Though this direction is nearly the same as that of the furrows offshore, there is the suggestion that these are debris laden water bands driven from the coast to the ocean, and which are also laden with sediment eroded from the beach and littoral by wave action. Its fingerlike plumes have been sheared at their seaward end by a current flowing southwestward parallel to the coast, as it appears on the undermost part of the photographs (colour infrared (A) and black and white panchromatic (B). Their orientation indicates that the velocity of the current increases in an offshore direction from a distance of 10 to 20 km off the coast. This interpretation coincidences with that of R. E. Hunter, 1973, concerning ERTS imagery of this area.

The dark blue strip offshore (colour infrared A) may indicate a current parallel to the coast. Currents like this are called "coastal jets" by some oceanographers. Certain features of the system of marine currents on the continental shelf can be deduced from satellite photographs or images such as these.

Indications of the bottom topography give a pattern of bands parallel to the beach and coast to a distance offshore of 15 km, where a depth of 12 m is found. Driving forces are waves and currents. Further offshore the pattern of bands is somewhat similar to the interpreted drift pattern, so there is no certain identification possible from this satellite photograph.

In the area of the entrance to Galveston Bay, around the jetties, both phenomena – submarine bottom features and turbid water masses – are certainly present.

The bottom is composed of sand, mud and clay which is receiving sufficient mechanical stirring from the general water circulation and waves to present a turbid appearance on the satellite photograph. It can therefore be seen that the main difficulty is to distinguish between the bottom topography and the current patterns.

Moreover, note that the pattern of clouds, visible on the colour photograph, has a similar pattern; this is a truly atmospheric phenomenon. This apparent water discolouration extends almost 15 km off the coast, but it has been observed that discoloured water from the lagoons can move to the edge of the continental shelf, i. e. more than 100 km offshore. Furthermore on the infrared colour photograph (A), there are some plumes of turbid waters more than 40 km off the coast.

Evaluation of orbital remote sensing with multispectral photography

The interpretation of this series of satellite photographs can help to detect how far the water interaction of a lagoon (here Galveston Bay) reaches into the ocean, and gives an indication of the pattern of sediment transport with currents nearshore and offshore.

Coastal boundaries surveys

Multispectral orbital photography is a tool for coastal boundary survey.

Water is imaged as almost pure black on the infrared photograph, whereas the terrestrial or non covered beach occurs in light grey or white.

Due to this effect the separation of beach and sea is depicted with high contrast and can be used for coastal mapping, especially for coastal boundary surveys used for rights of ownership. This is important, since the coastal zones of the earth are being given new emphasis due to the importance of the exploitation of the mineral resources and fisheries, therefrom.

Bibliography

Hunter, R. E., 1973, Distribution and movement of suspended sediment in the Gulf of Mexico off the Texas coast: NASA Goddard Space Flight Center, Symposium on Significant Results Obtained from the Earth Resources Technology Satellite-1, 2d, New Carrollton, Md., Mar. 1973, Proc., vol. 1, sec. B, pp. 1341–1348.

Leblanc R. J. and W. D. Hodgson, 1959, Origin and development of the Texas shoreline: 2nd Coastal Geography Conference, Louisiana.
And: see chapter "Houston-Galveston".

Comparison of satellite colour photography and ERTS–1 MSS imagery concerning features and patterns of inter-tidal zone (tidal flats, mangrove vegetation) and phenomena and processes (change detection) of estuarine and marine zones (turbid water masses, tidal streamflow dynamics, interface of estuarine and oceanic waters and rip currents)

Golfo de Fonseca, coast of Central America, Pacific Ocean, bays, estuaries and lagoons

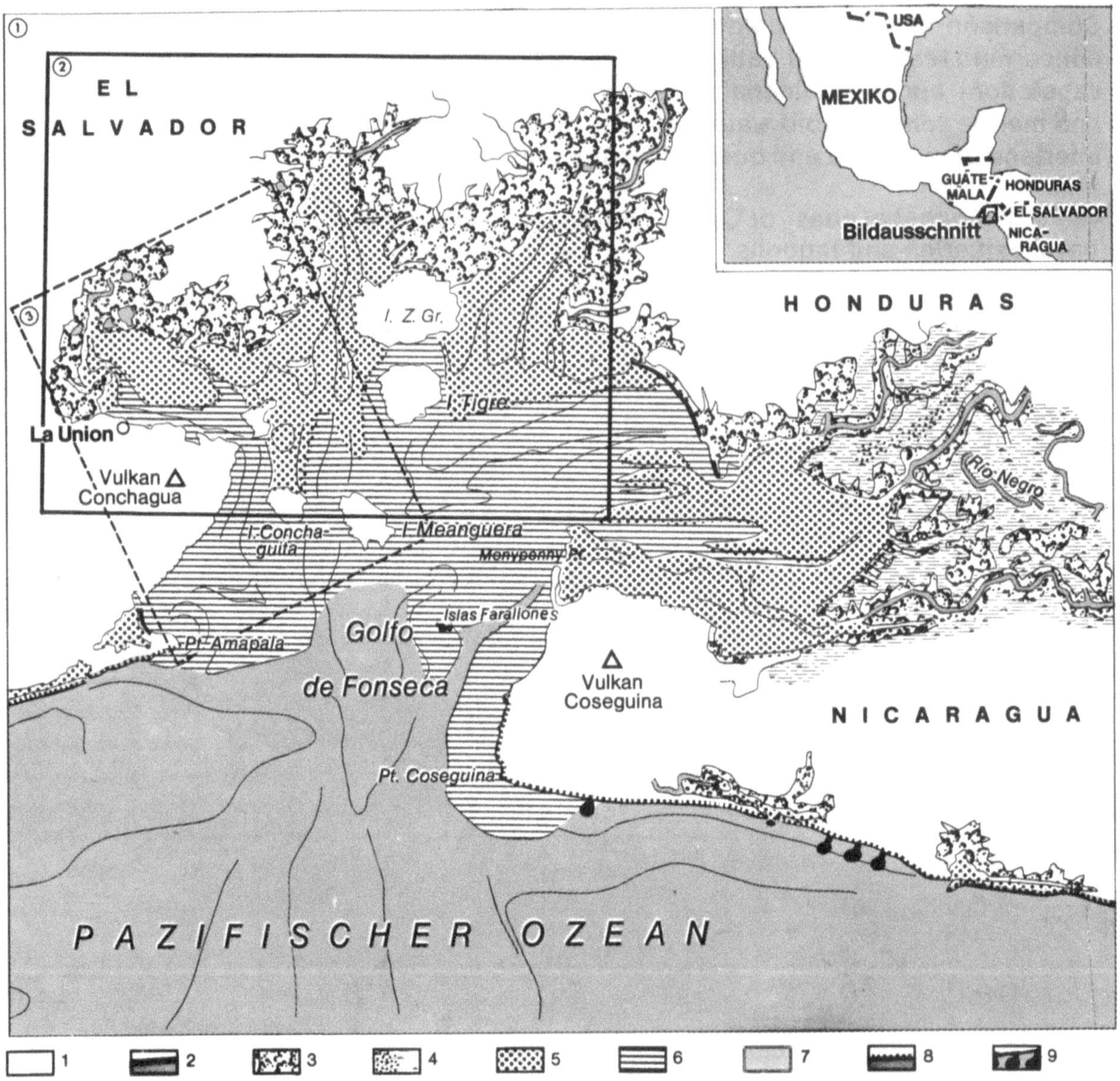

I. Z. Gr. = Isla Zacate Grande

Golfo de Fonseca, Coast of Central America, Pacific Ocean Bays, estuaries and lagoons

Interpretation map, from NASA satellite colour photograph

APOLLO mission 9, no. AS 9-19-3019

Interpreter H. G. Gierloff-Emden, 1975

Legend of topographic and hydrographic features classified in zones with selected discriminated objects (generalized: no differentiation of distinct water levels on this small-scale map)

Scale approximately 1 : 660 000, slightly oblique photograph

1 Frame of part of Apollo satellite photograph (interpretation map)

2 Frame of ERTS satellite MSS image (section) western region of the gulf

3 Frame of aerial photograph

4 Frame of ERTS satellite MSS image (section) eastern region of the gulf (4 is not marked as a frame in this interpretation map)

(1, 2, 3, 4 compare particular photographs and MSS images)

I above High Water level: terrestrial

mainland on the map as white area

1 mainland, prevailing mountainous

2 sandy beaches on barrier islands, beach ridges and spits

II between High and Low Water level: inter-tidal

3 mangrove, vegetation

4 tidal flats in mangrove areas

5 tidal flats with half-tide channels (on inter-tidal ground) partly covered from turbid water

III below Low Water level: estuaries and marine

5 turbid water masses of bays, estuaries and lagoons (grayish-red colour on printed satellite photograph) overlaying tidal flats and submarine shallows

6 slightly turbid water masses of middle gulf (grayish-blue and blue-green on printed satellite photograph)

7 oceanic water masses (purplish-blue, blue and greenish-blue colour on printed satellite photograph)

8 breakers (hydrographic phenomenon of scale of width of some 50 m to 100 m extension in the surf belt) on fore-shore (white-grey colour on printed satellite photograph); spilling breakers on the coast west of Punta Amapala occur as broad ribbon, visibility depending on foamy patches; surging breakers on the steep coast of Punta Cosegüina occur as more narrow ribbon)

9 rip currents (hydrographic phenomenon of scale of some 100 m in length, on the map a little exaggerated in comparison to satellite photograph) (light grayish-blue colour on printed satellite photograph)

Satellite photograph:

Golfo de Fonseca, Coast of Central America, Pacific Ocean, bays, estuaries and lagoons

Flight information data:

Mission: *NASA – APOLLO 9*

Date: *9. March 1969*

Photograph: *colour*

Magazine: *A frame no. 3019*

Camera: *Hasselblad 500 C*

Lens: *Zeiss Planar f/2,8; 80 mm*

Film: *Kodak Ektachrome*

Filter: *haze*

Altitude above mean sea level: *101 naut. miles*

Geographical position of the principal point:
latitude: *12 degrees 52 min. N*
longitude: *87 degrees 45 min. W*

Chart: *ONC – WAC code K – 25*

Local observation time: *15.15*

Sun azimuth: *225°*

Sun elevation: *38°*

Photograph viewing mode: *slightly oblique, but near vertical*

Cloud cover: *3 % alto cumulus*

Scale of the printed picture: *1 : 660 000 (approximate)*
(section of the satellite photograph)

NASA – APOLLO 9

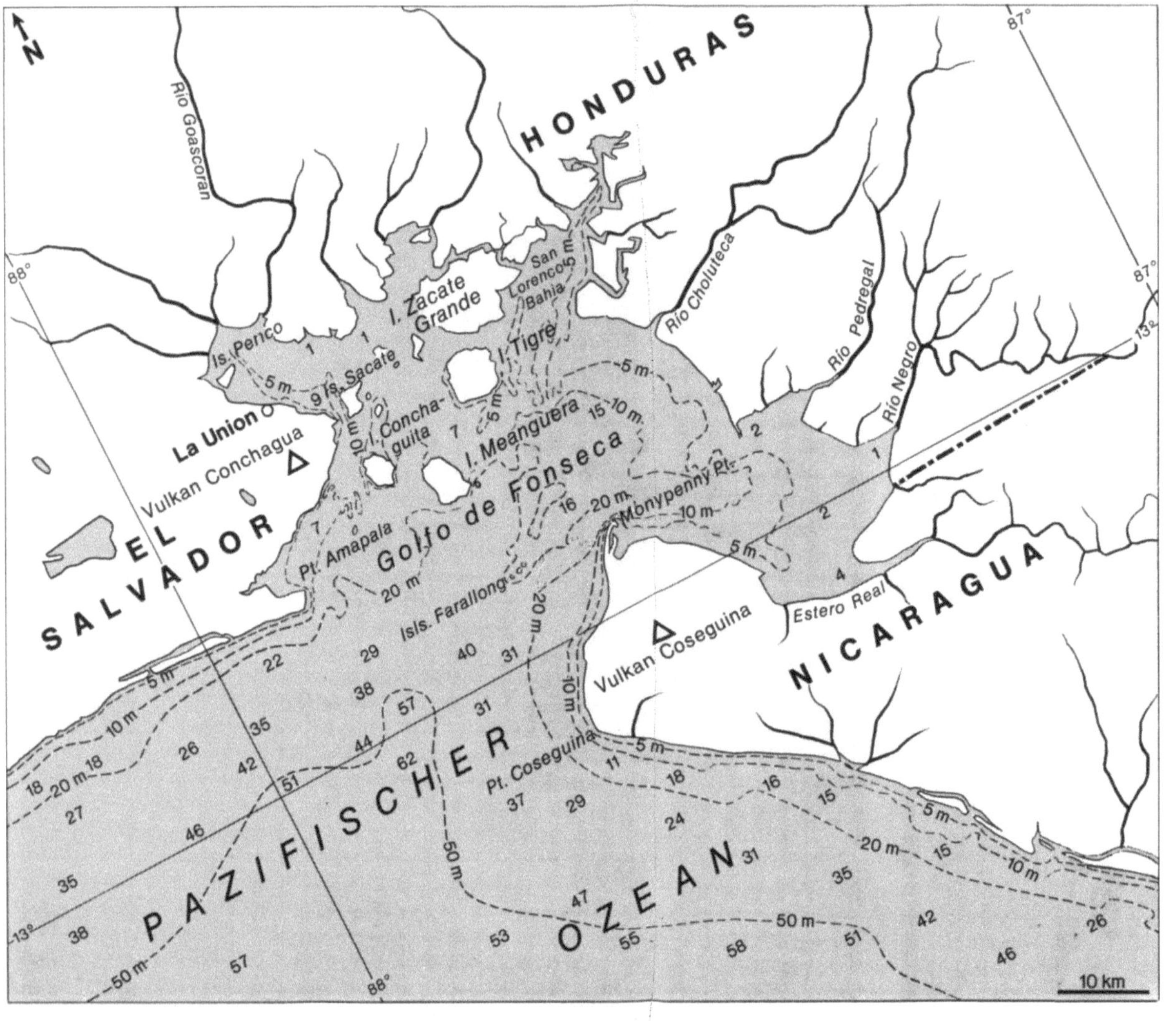

Golfo de Fonseca, coast of Central America

Chart of the area of the satellite photograph, scale approximately 1 : 750 000, i. e. close to the scale of the printed Apollo 9 photograph which is slightly oblique, i. e. which has a variable scale with bathymetric topography

Comparison of satellite colour photography and ERTS–1 MSS imagery concerning features and patterns of inter-tidal zone (tidal flats, mangrove vegetation) and phenomena and processes (change detection) of estuarine and marine zones (turbid water masses, tidal streamflow dynamics, interface of estuarine and oceanic waters and rip currents)

Golfo de Fonseca, coast of Central America, Pacific Ocean, bays, estuaries and lagoons

The following study is a presentation of the use of satellite colour photography (NASA Apollo mission) and satellite scanner imagery (NASA ERTS–1, since 1975 renamed "Landsat–1").
The area selected is on the coast of Central America – on the Pacific Ocean – the "Golfo de Fonseca". This area is fairly well known to the author, who did some fieldwork there in 1954 and 1955. Literature: H. G. Gierloff-Emden, 1959, Die Küste von El Salvador, eine morphologisch-ozeanographische Monographie.
This is an important method of studying comparative features using multiband imagery to consider wetland resources. There are differences in the spectral reflectance of wetland scenes, so it is necessary to find the optimum remote sensing media for differentiation and comparison.
The purpose of this paper is to examine the use of such satellite pictures and imageries for boundary mapping of: shorelines and wetlands with tidal flats and mangroves; inter-tidal water movements with translation of turbid water as interchange between estuary, lagoon or bay and the ocean; and some geomorphological features.
In comparison with planned studies for such purposes, as carried out with a sequence of multiple flights taking into account the unit time (time scale) of tidal rhythm, with the successive water filling of tidal areas from low water to high water, this is the use of satellite pictures taken at only one instant of time, depending on the satellite orbital parameters.
The large time scale of unit time of climatic conditions, such as seasons of plant live, is not of importance for the study of this selected region, because the vegetation is a type of mangrove, which is evergreen all the year around and does not have specific seasons of plant primary production, so that there would be no difference in spectral reflectance the year around. Moreover, both the Apollo satellite photograph (March) and the ERTS–1 MSS imagery (Dec. and Feb.) were taken during the dry season of the study area.

Interpretation of the colour satellite photograph, Golfo de Fonseca

Water masses

The dynamic nature of turbid water masses is well illustrated. Interpreting this colour satellite photograph one can identify three water masses indicated by colour and distribution (numbers as used in the legend on the interpretation map):

III 5 turbid water masses (grayish-red colour) near the inner bays and estuaries,

III 6 slightly turbid water masses (grayish-blue colour) in the middle of the Golfo de Fonseca,

III 7 oceanic water masses (purplish-blue, blue and greenish-blue colour).

The turbid water masses (5) cover the bottom of the inner Gulf. Since they are opaque no indication of submarine topography or water depth can be obtained from the photograph in these covered areas.
Small, long bands of grayish-blue waters (6) and of blue oceanic waters (7) extend into the Golfo de Fonseca. It is extremely difficult to decide at which moment of the oscillating tidal process the photograph was taken. As it is known by the author from personal experience (he once travelled around the Gulf), i. e. so called "pre-information", the water masses have well marked sharp edges when full flood or ebb tide streams are running (this cannot be observed on the photograph). It is probably the time close to low tide, because the flats in the inter-tidal regions around the inner bay and mangroves appear to be partly exposed or to have fallen dry. (Description in the NASA catalogue states: probably beginning of incoming tide).
But, anyway, three water masses in the Gulf can be detected. Translating this configuration into the time scale of tidal processes (as mentioned above), the conclusion seems to be certain, that these water masses exist for more than one tidal cycle and mix only over several semidiurnal periods of days or perhaps a week. This is of great importance to both, the fishery and pollution activities in the Gulf, and the satellite photograph can help to map the limits of the water masses simultaneously.

Breakers and rip currents

Two more hydrographic phenomena of this area can be observed in this satellite photograph: both are of small size and are very close to the minimum visible in this satellite photograph; also both are of small scale in units of time, as are the processes which produce them. (Numbers as used in legend of interpretation map).

III 8 A very marked ground swell exists off the coast, causing breakers and a surf belt along the shore of the Pacific Ocean. The heavy breakers along the Pacific coast occur as a small white line with small dots on the photograph. These breakers are some 50 to 500 m long, and with a width of 50 to 100 m, running on to the beach in a crescentic pattern and in time units have periods of 10 to 25 seconds. (Spilling breakers on the coast west of Punta Amapala occur as a broad ribbon, depending on foamy patches. Surging breakers occur on the steep coast of Punta Cosegüina as a narrower ribbon).

III 9 The distribution and extension of rip currents along the southern coast of Nicaragua (light grayish-blue colour on satellite photograph). These rip currents have lengths of 50 to 400 m and represent undertow currents which exist as distinct objects with a time unit of about 1 to 10 minutes.

Some results

To obtain information on the phenomenon of interchange of water masses between bays, lagoons or estuaries of the gulf and ocean, due to water transparency and the pattern of turbid water masses, both colour photography from satellites and ERTS-MSS imagery, seem to have some advantage. The colour film emulsion used gives a fairly good indication for this purpose, as do the ERTS–1 scanned bands, especially no. 4 and no. 6, which have some differentiated penetration through the water. Both imagery types can indicate the discolouration of turbid water masses.
The colour satellite photography shows up a more differentiated distribution of turbid tidal stream waters, alternating with some oceanic waters in a pattern of bands, stripes, furrows and plumes in the middle region of the gulf, in comparison to the ERTS-MSS images.
But the ERTS-MSS image, especially the MSS-band no. 5, indicates the boundary of the turbid gulf water mass against the oceanic water mass at the entrance of the gulf as the interface of two entirely different

water masses. Moreover, this ERTS-MSS band indicates the extension of gulf water masses of the former ebb tidal stream cycle into the ocean, to a distance of 30 km to 40 km; i. e. the detectable discolouration of gulf water mass in the Pacific Ocean as an interchange area (compare ERTS–1 MSS band from March 1973).

But for discrimination of boundaries of various features as for vegetation, such as the mangrove areas, for water surfaces and for dried sandy or muddy tidal flats, or for delineation of shorelines, the scanner imagery of ERTS–I with four spectral bands delineates the boundaries far better than the colour satellite photograph.

The boundary water against any other feature shows up in sharp contrast on the ERTS-MSS band 7, the infrared. The original copies of NASA colour photographs and ERTS-MSS scanner imageries are of much higher quality than the prints presented in this study are or can hope to be.

Both examples prove how important satellite photographs and imageries can be for small scale mapping of wetlands and coasts.

Conclusion and comparison of colour satellite photograph with ERTS-MSS images

The north-western region of the Golfo de Fonseca is covered by the ERTS satellite MSS 4-band images. The interpretation of this area (see following pages) provides an opportunity for comparison with the above Apollo colour photograph:

On the ERTS images, mangrove vegetation and the water line can be mapped with much greater accuracy, due to the different bands, each with special spectral sensitivity, with the ERTS-MSS band 7, however it seems that in detecting the water masses in the Gulf, the Apollo colour photographs have advantages in comparison with images of single bands of the ERTS-MSS system.

But colour composites of 3 bands of the ERTS-MSS images, as published by the Department of the Interior, US Geological Survey, 1972, "ERTS-MSS images of the region of the eastern Golfo de Fonseca, taken on December 24, 1972", printed on the same scale as the Apollo photograph, gives an indication of water masses of much the same quality as the Apollo colour photograph in that specific area of the Gulf. ERTS-MSS images give however a brilliant separation of mangrove vegetation and flooded tidal flats or half tidal channels on inter-tidal ground, because on these colour composite images mangrove vegetation comes out in red (ERTS-MSS band 7) and the tidal channels in blue, i. e. in high contrast.

Interface and boundary of water masses in the Gulf of Fonseca

Optically well marked boundary between water masses, caused by a tidal stream. There is a visual difference on the surface concerning discolouration and roughness, causing different reflecting conditions.
Water masses and boundaries: compare satellite photograph and image. View looking northward, taken from the entrance. Coast on the left: mainland; coast on the right: Isla Conchagüita.
Photograph taken by H. G. Gierloff-Emden, 1955.

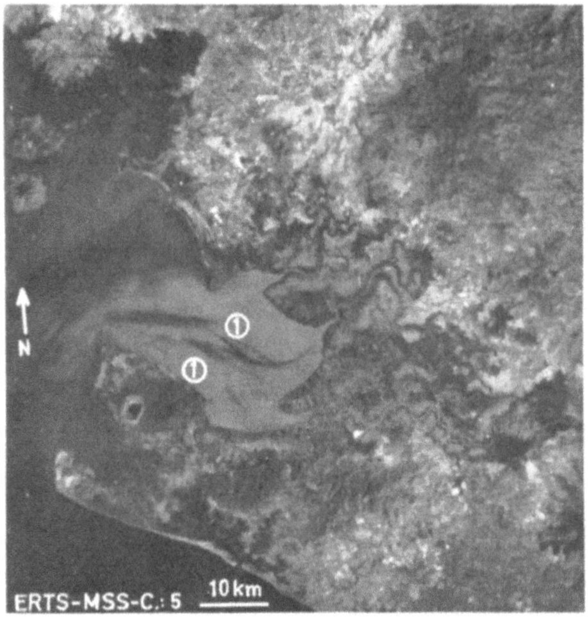

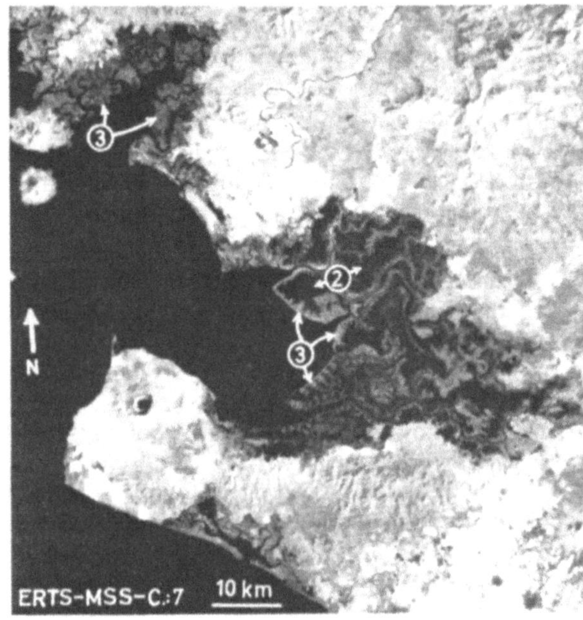

NASA ERTS–1 satellite MSS imagery

Geographical location of the scene:

Golfo de Fonseca, Coast of Central America (eastern region of the Gulf)

Images: *MSS band 5, left figure and band 7, right figure*

NASA image material: *black and white 2,2 inch film positive, scale 1 : 3 369 000 for each band*

Printed picture: *reproduction, enlarged, of NASA image material, section of the MSS scene*
Note: the printed picture of this publication is autotype and bears less details and pictorial information than the NASA film copies

ERTS–1 scene: *ID 1154-1538-4 to 7*

Date of acquisition: *24. Dec. 1972*

Local observation time: *09. 30.*

Sun azimuth: *139°*

Sun elevation: *42°*

Image viewing mode: *vertical*

Scale of printed pictures: *1 : 1 000 000*

Interpretation of ERTS–1 MSS imagery

State of tide at time of data acquisition is close to High Water

Left picture: ERTS MSS band 5

1 turbid water masses (light grey) covering tidal flats and submarine shallows of the bay

Right picture: ERTS MSS band 7

2 tidal flats in mangrove areas covered at time of data acquisition (black)

3 mangrove vegetation (grayish white)

Comparison with Apollo 9 colour satellite photograph:

Detection, identification and differentiation of these features: (1) is ambiguous on the Apollo-photograph, because in very shallow areas, the bottom may be visible. This is not possible on ERTS-MSS-band 7 due to lack of penetration

Detection, identification and differentiation of these features: 2 and 3 sharp boundaries on ERTS image

3 additional: sharp boundary between vegetation features and water;
no pictorial information can be gained for tidal flats due to infrared photography (water appears black)

Bibliography

Gierloff-Emden, H. G., 1959, Die Küste von El Salvador. Eine morphologisch-ozeanographische Monographie: *Acta Humboldtiana*, Wiesbaden.

Krügler, F., 1966, Über eine optisch markante Stromgrenze an der Polarfront in der Dänemarkstraße: DHZ, vol. 19, no. 4, p. 159–170.

Römer, E., 1936, Örtlich-periodisch auftretende Kabbelungen an der mexikanischen und mittelamerikanischen Westküste: Ann. Hydrogr. mar. Meteorol., Spec. Pap. of Sept. no., p. 55.

West Coast of Central America and United States Pilot: Hydrographic Department, Admiralty, London.

Maps

Chart ONC – WAC code K – 25

International map ND-16 Tegucigalpa, in scale 1 : 1 000 000.

Nautical charts, U. S. Nav. chart, H. O. secr. of Navy no. 0973, Gulf of Fonseca, in scale 1 : 145 278.

British Admiralty charts 1049, 587, 786.

Interpretation of ERTS—MSS images, Golfo de Fonseca

The north-western region of the Gulf

NASA ERTS–1 satellite MSS imagery

Geographical location of the scene:
Golfo de Fonseca, coast of Central America (northwestern region of the Gulf)

Geographical information: *drafted in image band 7*

Some pictorial information: *drafted in image band 7 and band 5*

Images: *MSS bands 4, 5, 6, 7*

NASA image material: *black and white, 7,3 inch film positive scale 1 : 1 000 000*

Printed picture: *reproduction and enlarged amplification of NASA image material, section of the MSS scene note: the printed picture of this publication is an autotype which gives less detail and pictorial information than NASA film copies*

ERTS–1 MSS scene: *ID 1209-1544 to 47*

Date of acquisition: *17. Feb. 1973*

Local observation time: *09.00*

Sun azimuth: *123°*

Sun elevation: *40°*

Image viewing mode: *vertical*

Scale of printed pictures: *1 : 325 000*
remark: the scale of printed ERTS imagery is about twice that of the printed Apollo colour photographs in this publication

Interpretation of ERTS–1 MSS imagery

See interpretation key and the area within frame 2 of the interpretation map and text of this publication of the Apollo 9 satellite colour photograph

Note: enlargement of NASA film copies up to three times of these printed pictures presents a better visibility of features and patterns to the human eye without magnifying instruments, but as noted above the printed picture is an autotype, not a photographic emulsion picture, and it therefore gives less detail and pictorial information than NASA film copies

General interpretation remark: ERTS–1 MSS imagery presents differentiation of tidal mud flats and turbid water on MSS band 5, and sharp boundaries of terrestrial features and mangrove vegetation against water on band 7.
For details of features and patterns compare the aerial photograph in this publication in area of frame 3 of the interpretation map which shows Bahia Puerto La Union (Port La Union Bay). Tidal flats, mostly uncovered within the mangrove area are visible on ERTS MSS bands 6 and 7 but not a detailed pattern as on the aerial photograph.
The image acquisition time is coincident with tidal High Water: band 7 shows (water occurs black) that tidal flats are covered (compare band 5) and tidal channels are filled

ERTS-MSS-C.:4

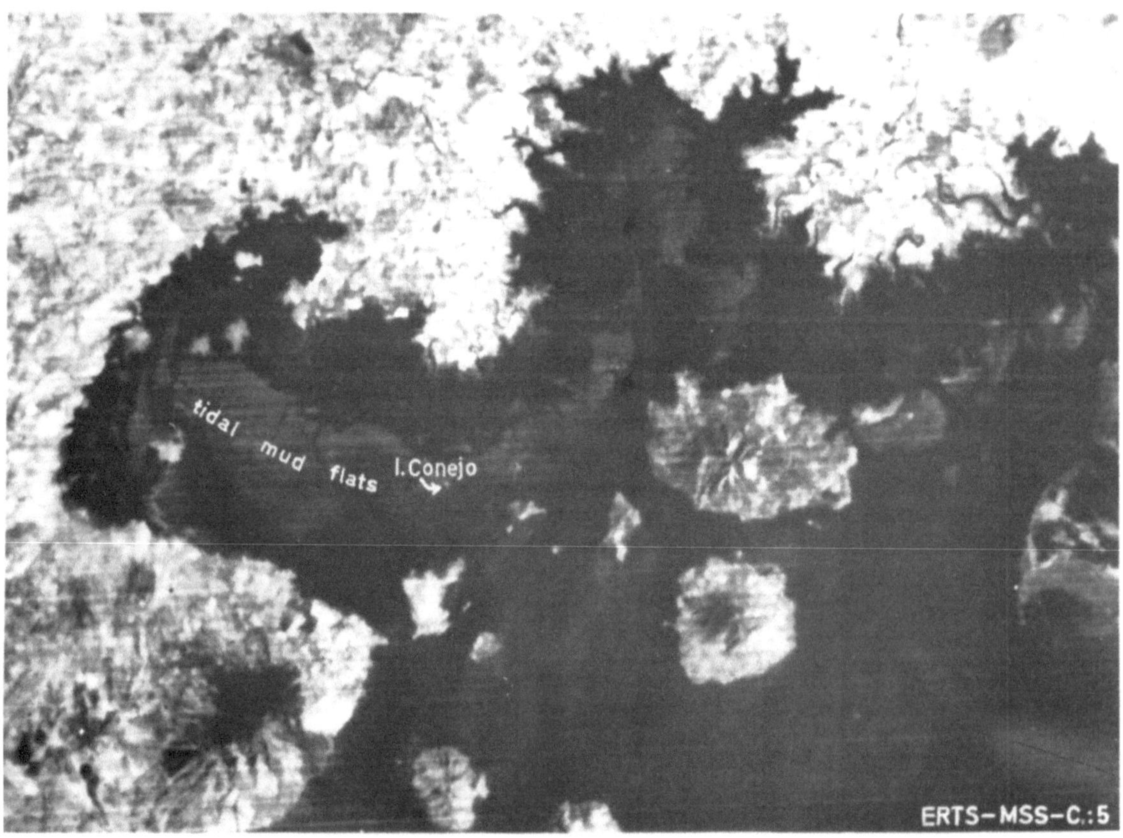
tidal mud flats
I.Conejo
ERTS-MSS-C.:5

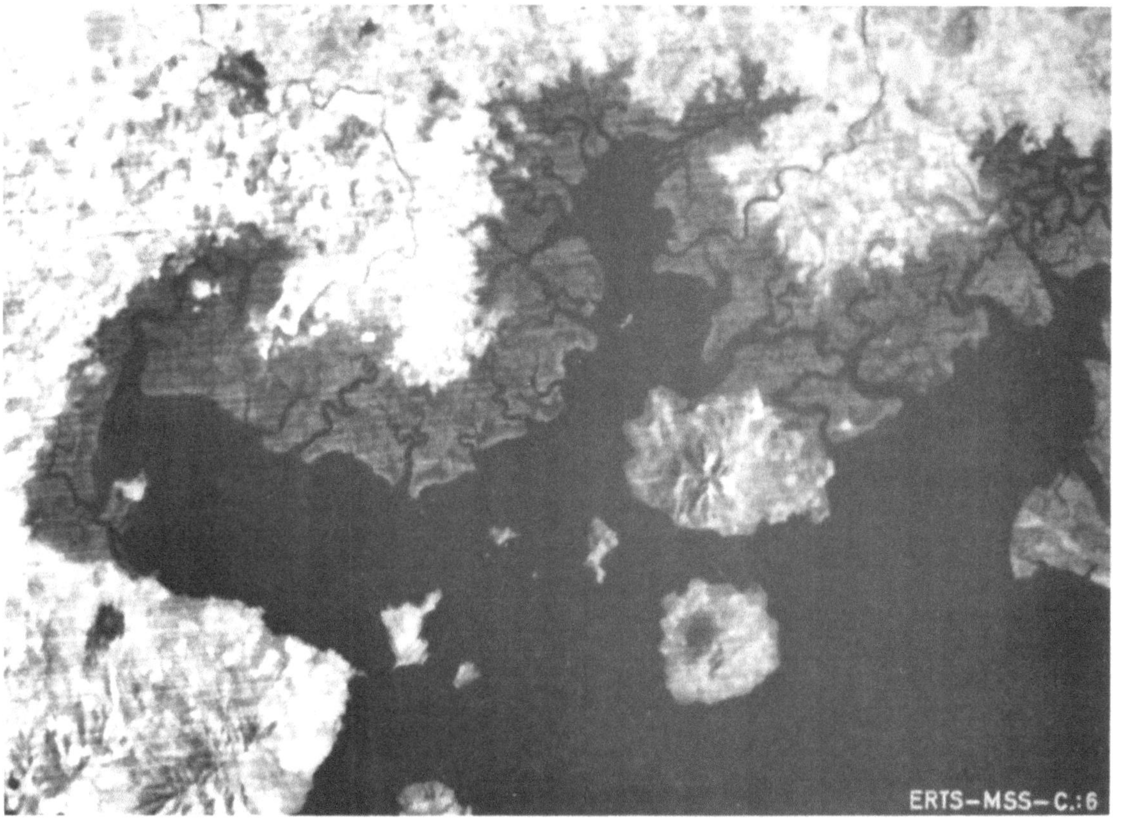
ERTS-MSS-C.:6

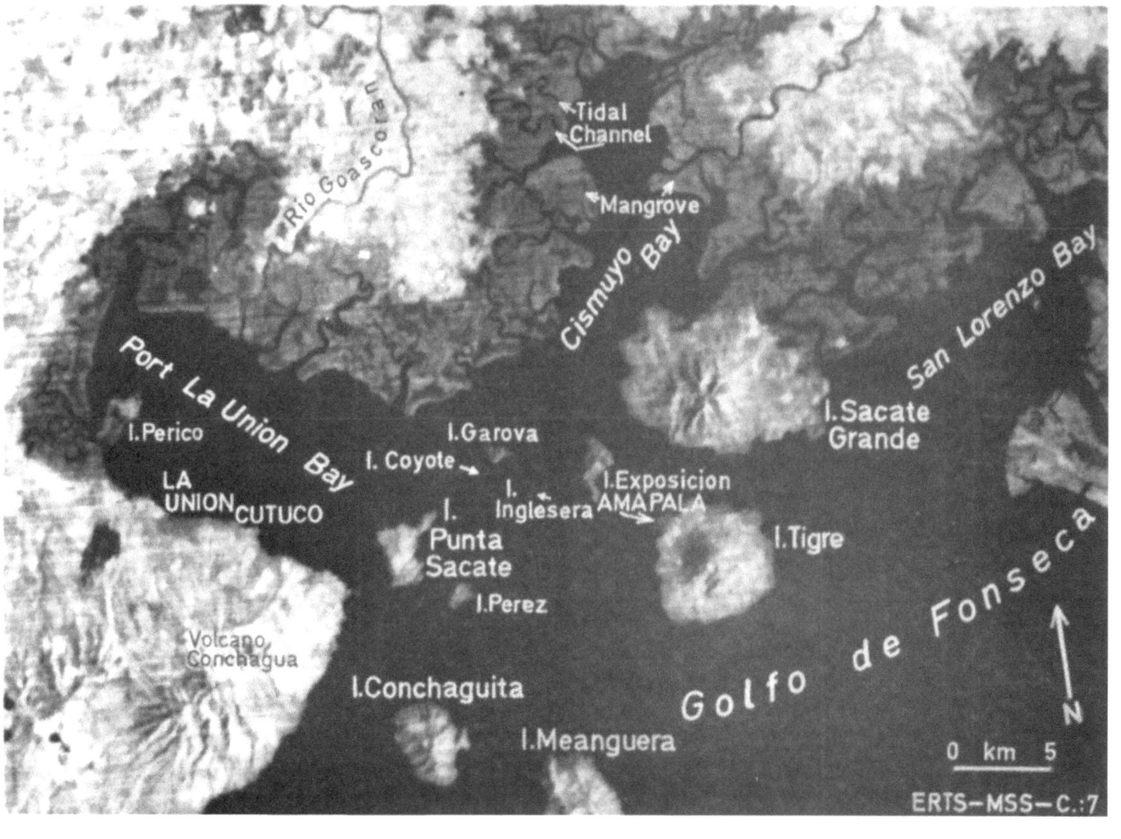
Tidal
Channel
Rio Goascoran
Mangrove
Cismuyo Bay
San Lorenzo Bay
Port La Union Bay
I.Perico
I.Garova
I. Coyote
LA
UNION
CUTUCO
I.
Inglesera
I.Exposicion
AMAPALA
I.Sacate
Grande
I.
Punta
Sacate
I.Tigre
I.Perez
Volcano
Conchagua
I.Conchaguita
Golfo de Fonseca
N
I.Meanguera
0 km 5
ERTS-MSS-C.:7

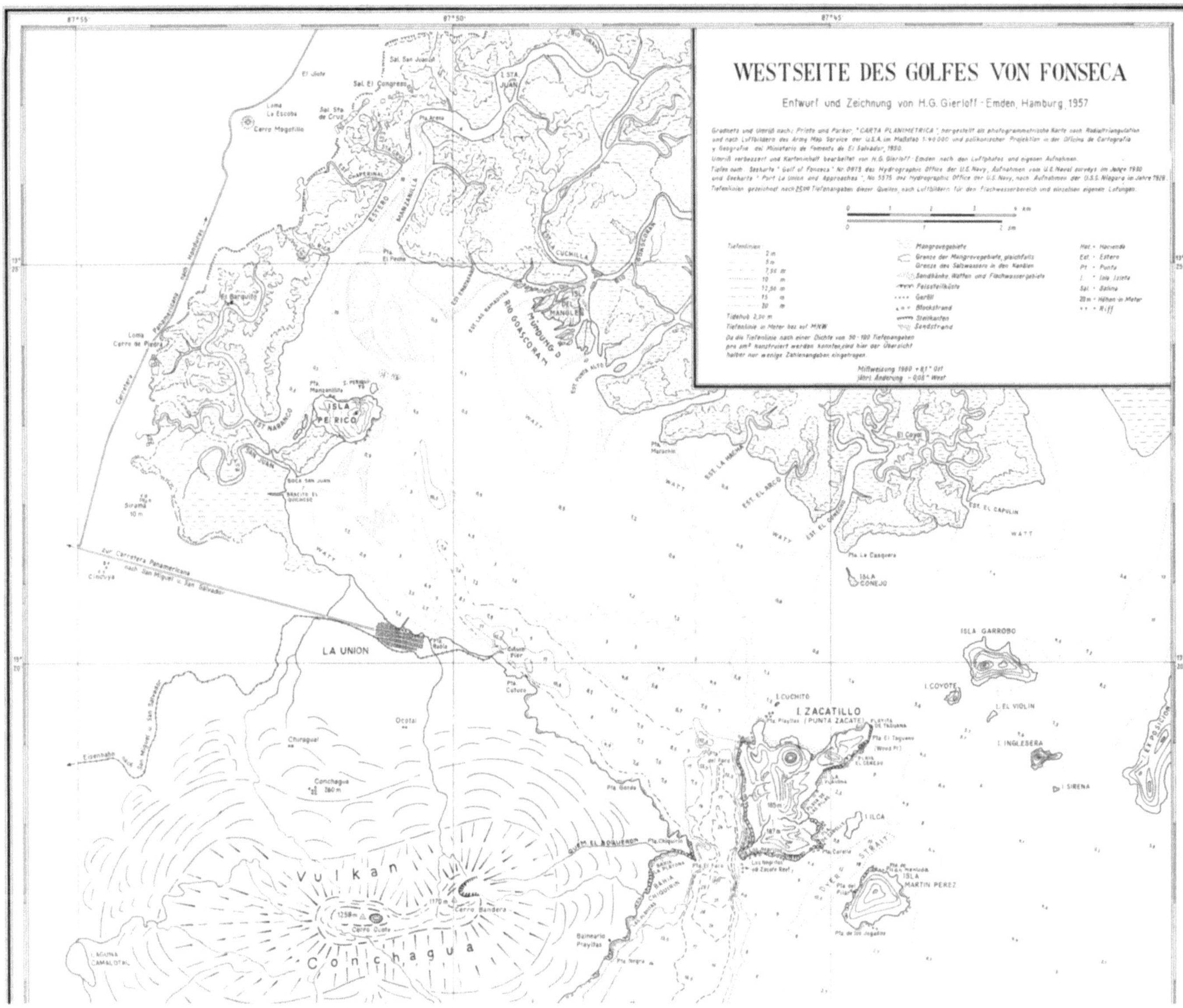

WESTSEITE DES GOLFES VON FONSECA
Entwurf und Zeichnung von H.G. Gierloff - Emden, Hamburg, 1957
Mangrovegebiete
Sandbänke, Watten und Flachwassergebiete
LA UNION
ISLA PERICO
I. ZACATILLO
ISLA GARROBO
ISLA CONEJO
I. COYOTE
I. INGLESERA
I. SIRENA
I. ILCA
I. CUCHITO
ISLA MARTIN PEREZ
Vulkan Conchagua
Cerro Bandera
El Coyol
WATT

Interpretation of ERTS—MSS images, Golfo de Fonseca

The north-western region of the Gulf

The landscape levels will be analyzed according to the MSS-bands for that part of the Gulf of Fonseca. The western and north-western part of the Gulf was mapped by the author in 1954–1955. Gierloff-Emden (1958).

I The terrestrial zone

Image interpretation: Isla Conejo (diameter 100 m) is detectable on MSS-bands 5 and 6, Isla Coyote (diameter 300 m) on MSS-bands 4, 5 and 6. This illustrates the better resolution (minimum visible) of the MSS-band 6 compared to the MSS-band 7. For small islands in the sea this is the resolution limit-ground resolution and image resolution (depending on contrast).
The coasts surrounding the volcano Conchagua are steep and rocky. The land-water border is sharp and distinct on MSS-bands 5, 6 and 7, whereas MSS-band 4 delivers at some locations an ambiguous information.

Environmental features: In the inner western and northern corner of the bay tidal flats and mangrove occur. The inner border of the tidal landscape against the land is distinctly shown by the relief. The land begins with a 20 m high escarpment at whose foot several salt pans are situated.
The mangrove belt ends at the northern slope foot of the volcano Conchagua 2 km west of La Unión.
The harbour of La Unión lies 2 km east of the city of La Unión and is called Cutuco.

Image interpretation: (MSS-band 7: land bright grey, tidal landscape middle grey). The urbanized region of La Unión shows up well on MSS-band 7, less well on MSS-band 6. The 200 m long and 30 m wide landing bridge, built into the bay is below minimum visible and not detectable.

II The inter-tidal zone

Environmental features: Extensive mangrove areas, which are interlaced with many tidal channels, surround the north-west bay of the Gulf of Fonseca. They reach their largest extension in the delta of the Rio Goascorán (as shown on the air photos and as detected by ground check).
The river, which is the boundary for 40 km between El Salvador and Honduras, meanders in the last part of its lower course and through its delta and estuary enters into the Gulf of Fonseca. The main junction arm was distinguished in 1954 by a 2 km long delta shaped form, which was built up of various flat sand and mangrove islands. The whole delta area is over 16 km wide. Due to its position in a tidal bay, it has been formed by the accumulation of debris from the Rio Goascorán and the tidal stream process.

Image interpretation: This delta is, as shown on ERTS-satellite image (channel 5), substantially silted up and overgrown since 1949 (comparison with aerial photograph of the year 1949).

Environmental features: The mangrove stands border the various channels of the inner bay of the Gulf, but are lacking on the large central areas of the flats. These open, higher and temporarily arid areas are mostly composed of sand, whereas the borders of the channels contain sandy mud and muddy substratum.

Image interpretation: These mangrove and tidal flat areas can be delineated well from MSS-band 7. The mangrove there appears light (infrared picture) against the damp sand. It can be established, from the form of the sand area, that no important changes have occurred since 1949 (comparison with map). In front of the mangrove covered border edge of the delta towards the bay lie extensive flats of which a small portion is built up of sands, but the greater part consists of one metre deep, soft mud, recognizable on MSS-band 5 but, better on MSS-band 4. On MSS-band 4 the submarine topography is covered and concealed with clouds of suspended matter in the waters (tidal inundation). Dead clam banks form more solid strips on the flats. These are not identifiable on the satellite image (below minimum visible). In the delta-region, especially in the eastern part, some cone shaped hills rise above the flat mangrove areas and flats; they are identifiable on all MSS-bands but show up better on MSS-bands 6 and 7.

Environmental features: The 10 m deep main water branch Estero Manzanillo threads through the mangrove region in wide loops. Its tidal estuary may have developed from a former mouth of the Rio Goascorán. The mangrove border is narrower on the north-west shore, and on the mainland begins with a sharply marked line.

Image interpretation: The deep channels of the river opposite the sand banks which are recognizable on the air photo, show only partly on MSS-band 5 on the ERTS-satellite-image.

III The marine zone (oceanic)

Environmental features: The 20 m depth contour of the ocean swings into the middle of the entrance of the Gulf, north to Meanguera island. The mean water depth to the north between the other islands is 6 m. The large water areas between the delta of the Rio Goascorán and the island Zacatillo are very flat, their depths are between 0 m to 1,50 m. The hydrographically most interesting phenomenon on the west coast of the Gulf is the deep channel which extends from Estero Manzanillo south along to Cutuco, the port of La Unión, cleared out by tidal stream. Through this passage the muddy water masses are led away at low-tide out of the channels of the delta area and the north-west bay. A very deep eroded channel lies between the Conchagua and the island Zacatillo, the greatest depth of which is a depression of 26 m depth with a maximum depth of 40 m. The surface current out of the passage reaches 3–4 knots. When the low tide from the passage moves southward, then the sharply defined boundaries between the blue sea water and the murky bay water, coloured yellow from sediments, is recognizable as far to the south as the latitude of Conchaguita. (On-site observation, ground check).

Image interpretation: These deeper submarine relief forms cannot be identified on any band of the ERTS-images; only sediment laden water bodies are discernable on MSS-band 4.

Tides and water masses

Environmental features: The mean tidal range on the Pacific Coast of El Salvador is locally enlarged in the large Bay of Fonseca and is 2,50 m; the mean spring range being 3,25 m to 3,75 m, and the mean neap range is 1,80 m. The salt content in the bay is oceanic and even in the north-west part does not fall below 30,5 ‰, i. e. in the water of the Gulf, the portion of ocean water dominates, the amount of fresh water inflow is small.

Image interpretation: Only bands 4 and 5 of the satellite images show differentiations in these oceanic features; demarcation of the water masses of the Gulf from those of the open sea is clearly identifiable. The murkiness of the Gulf water body appears as a light grey area in the Gulf and at its entrance. The waters of the Gulf are clearly distinguishable from the waters of the open ocean.

MSS-band 4 does not show any penetration because the murky water does not allow any recognition of submarine relief forms. MSS-band 5 shows a certain penetration of the water, but pictorial information is ambiguous.

Aerial photograph of Golfo de Fonseca, coast of Central America, northwestern region of the Gulf

covering Bahia Puerto La Union, estuary of Estero Manzanillo and Rio Goascorán delta and estuary with tidal flats and mangrove vegetation

Technical information

Aerial photograph: scale of original photograph *1 : 40 000*

Photograph viewing mode: *vertical*

Flight: *Army Map Service USA, 1949*
for Dirección de Cartografiá, San Salvador

Reproduction of aerial photograph to scale of printed picture: *1 : 60 000*

Interpretation: the dark edges along bay, estuary and tidal channels are areas covered with mangrove vegetation. The light areas in between mangrove vegetation margins are tidal flats, sandy areas, which are covered only at Mean High Water Springs (this is visible on ERTS MSS band 7).

The delta of the Rio Goascorán, as it appears on this photograph from the year 1949, is not the same or even similar to that on the ERTS-MSS imagery from the year 1973 (i.e. 24 years later). The delta has obviously changed in extension and pattern (compare text).

Use of multispectral scanner imagery of the NASA ERTS-1 satellite for interpretation and for small-scale mapping of environments on a tidal coast (tidal range values, coastal geomorphology, mangrove vegetation)

El Salvador, Pacific coast of Central America
The Lagoon of Jiquillisco (Spanish: Estero de Jiquilisco)

NASA-ERTS satellite MSS imagery

Geographical location of the scene:

Southeast region of the land and eastern region of the coast of El Salvador and Golfo de Fonseca, Central America

(Compare morphographic map)

Images: *MSS bands 4, 5, 6, 7*

NASA image material: *black and white, 7,3 inches film positive, scale 1 : 1 000 000*

Printed picture: *reproduction of a northern section of the MSS scene, on original scale 1 : 1 000 000*

ERTS–1 MSS scene: *ID 1209-1544 to 47*

Date of acquisition: *17. Feb. 1973*

Local observation time: *09.00*

Sun azimuth: *123°*

Sun elevation: *46°*

Principal point of image (as NASA code)
latitude: *13.070° N*
longitude: *88.405° W*

Image viewing mode: *vertical*

Scale of printed picture: *1 : 1 000 000*

Interpretation: see text and maps in the text

Note: mangrove vegetation and water-filled tidal channels at High Water Springs at data acquisition time

ERTS–MSS–C.:6

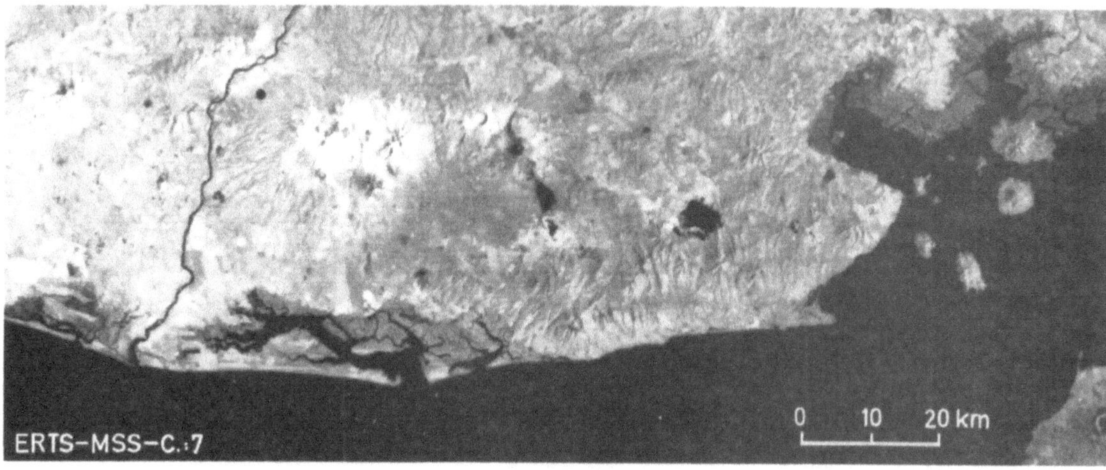
0 10 20 km
ERTS–MSS–C.:7

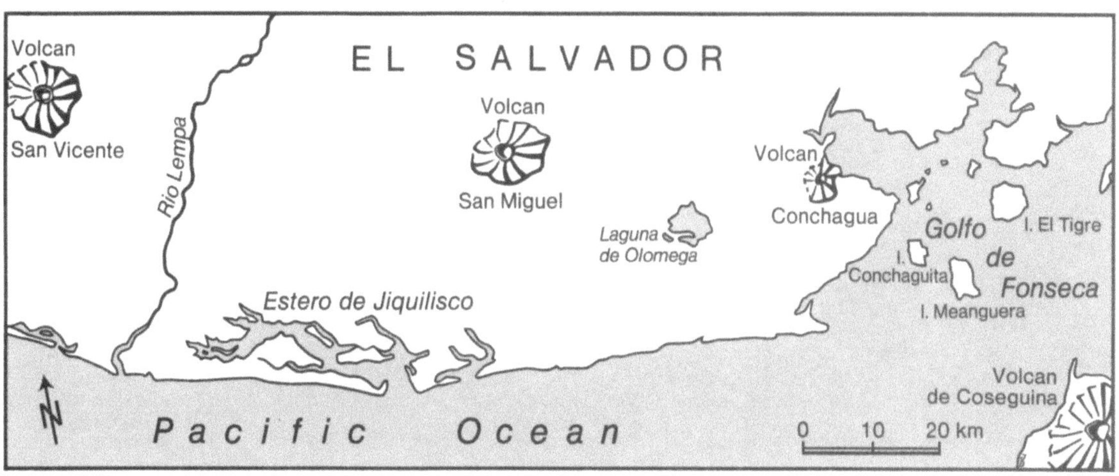
EL SALVADOR
Volcan
San Vicente
Rio Lempa
Volcan
San Miguel
Laguna
de Olomega
Volcan
Conchagua
Golfo
de
Fonseca
I. El Tigre
I.
Conchaguita
I. Meanguera
Estero de Jiquilisco
Pacific Ocean
N
0 10 20 km
Volcan
de Coseguina

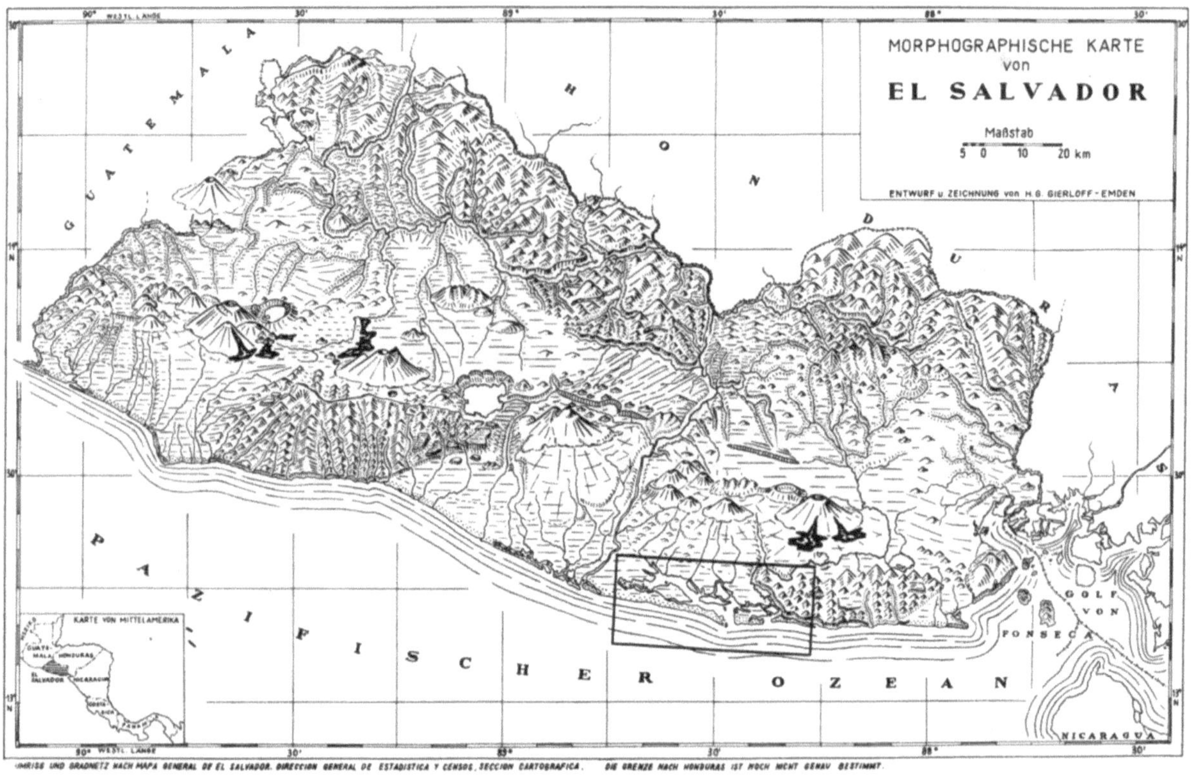

Morphographic map of El Salvador
Scale 1 : 1 800 000
compiled and drawn by H. G. Gierloff-Emden, 1958
region within small frame: Estero de Jiquilisco, region of a thematic mapping and interpretation of orbital ERTS–1 MSS imagery
compare also particular map

Use of multispectral scanner imagery of the NASA ERTS-1 satellite for interpretation and for small-scale mapping of environments on a tidal coast (tidal range values, coastal gemorphology, mangrove vegetation)

El Salvador, Pacific coast of Central America
The Lagoon of Jiquilisco (Spanish: Estero de Jiquilisco)

NASA ERTS–1 MSS 200 15/1088; observation: ID 1209-15444; date: 17. Feb. 1973;
NASA principal investigator: SR-331 d,–J. D. F 0452, H. G. Gierloff-Emden, NASA GSFC of 15. Feb. 1972.

Statement of the problem

In the present study the applicability of satellite MSS imagery to research and production of thematic maps on a scale of approximately 1 : 300 000 has been investigated for a subdiscipline of geography and oceanography, namely coastal geomorphology. The ERTS–1 MSS imagery has been used for the mapping of the coastline and the surface distribution of the inter-tidal landscape of a lagoon, the zones of tide water and the vegetation levels in the coastal environment.

Delimiting areas within a continuous surface on the map involves a decision process by the scientist. In the present study, the limits of the various specific habitats within the coastal zone were established on the basis of local knowledge of the environment including its features and processes, i. e. ground truth based on field work. H. G. Gierloff-Emden, 1959.

The boundary lines and boundary zones are investigated on the basis of their representation on the four MSS-bands of the ERTS-MSS imagery. Small features which are below the minimum resolution of satellite imagery, or which due to cartographic limitations cannot be shown on the map, have been disregarded.

Description of the area

The Estero de Jiquilisco forms a 50 km long section of the coast of El Salvador between the delta of the river Rio Lempa and the estuary of the river Rio Grande de San Miguel. The lagoon is affected by the ocean tides. The lagoon includes large tidal flats, mangrove areas, and water bodies and tidal channels. The lagoon is in some sense also an estuary.

The dimensions are:

length of the main arm	30	km
length of the large barrier peninsula San Juan del Gozo	30	km
length of the whole area	40	km
length of the whole area with Estero Espino	50	km
smallest width of the barrier peninsula	1,5	km
largest width of the barrier peninsula	3,8	km
open water areas of the main channel, greatest width	3,3	km
average width of the water channels	2	km
width of the entrance (Bocana San Juan)	2	km
landscape of the Estero from the barrier beach to the coastal plain	7 and 12	km
largest islands, maximum length	8	km
greatest water depth	20	m

The main channel of the lagoon, also called “Barra Santiago”, stretches from its western end in the general direction of WNW to ESE in smooth curves to the entrance (“La Bocana el Bajon”, i. e. “mouth”). The large barrier peninsula San Juan del Gozo separates the lagoon from the Pacific ocean.

By way of several channels, the lagoon of the Estero de Jiquilisco extends from the entrance Bocana el Bajon eastward to the estuary of the Rio Grande de San Miguel. It is connected by way of another channel with the Estero Espino which is the outermost easterly appendage of coastal lowland to the Jucuarán mountains.

The Estero region with its coastal lowland is bordered against the gently rising alluvial coastal plain by a low but distinct escarpment which appears as the edge against the mangrove vegetation.

A narrow strip of the vegetation type “lower evergreen forests”, mapped by W. Lauer, 1956, is surely present on this escarpment, though not recognizable in the ERTS images. The great alluvial plain of El Salvador is a sloping plain with a clearly visible incline of 1 : 1000 from N to S to the northerly edge of the Esteros.

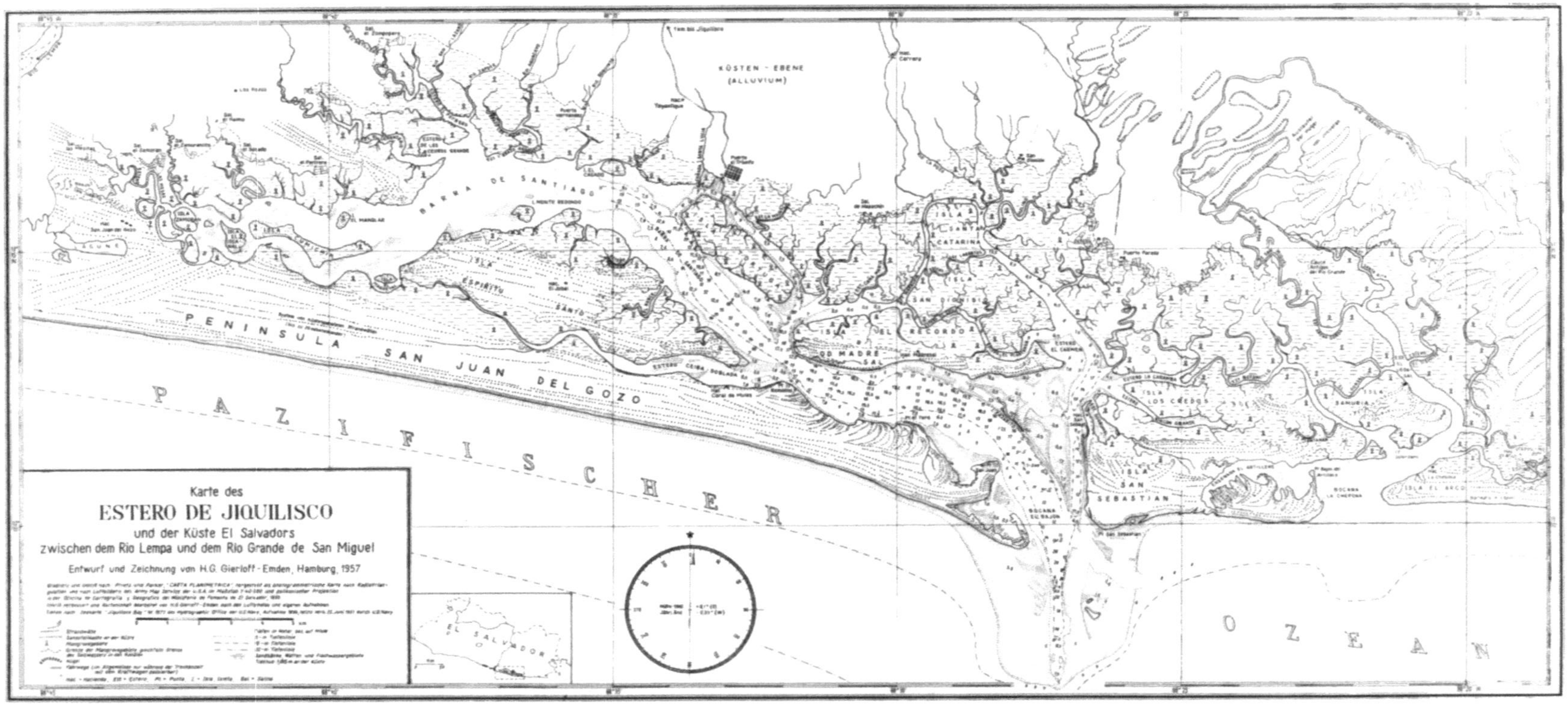

Map of Estero de Jiquilisco, Coast of El Salvador

Scale 1 : 210 000

compiled and drawn by H. G. Gierloff-Emden after fieldwork and aerial photographs. This map was produced by minimizing.

presenting: barrier peninsula and barrier islands, with beach ridges (dotted lines), mangrove vegetation and lagoonal water body, water tidal channels.

Compare the details of the tidal channels and of the mangrove vegetation of this map with the small scale thematic maps made from ERTS-MSS imagery, and the ERTS imagery itself.

See following pages.

Analysis, interpretation and mapping of the Lagoon of Jiquilisco, using ERTS–1 MSS imagery

I The terrestrial zone (above High Water level)

Features and pattern of this environment as elements of coastal geomorphology.

The coastline

Environmental features: The dry sand beach (the substratum is sand without stones and gravel) is the lowest seaward strip of the barrier beach with a width of 20–40 m. It is recognizable as a white strip on ERTS-MSS-bands 4 and 5 of the satellite image (compare figure).
Image interpretation: Due to its being a linear object, this narrow strip is still recognizable on ERTS-MSS-bands 4 and 5. Since ERTS-MSS-bands 4 and 5 provide water penetration to the white sand bottom of the foreshore, a part of the wet beach, covered at High Water, appears similar to the dry beach (which is above High Water). Thus, a wider strip as only the dry beach is visible, including narrow strips of two levels at the boundary of land and sea. On ERTS-MSS-band 7 the coastline occurs as a sharp defined boundary land – sea.

The coastal barrier complex

Environmental features: The beach ridges do not always run exactly parallel to the beach line. On the middle strip of the barrier peninsula large areas are used by burning the vegetation to gain grazing land for cattle and for growing maize.
Image interpretation: White strips are recognizable, especially on ERTS-MSS-band 5, land use pattern can be detected.

Environmental features: The barrier peninsula has bush and tree stands among which the coconut palm, the low, thorny carbon (Mimosa tenuiflora), the palm (Bactris subglobosa) called Huiscoyol, and amongst others several leguminocean plants occur.
Image interpretation: Dark strips are visible (especially on ERTS-MSS-bands 5 and 6).

Environmental features: The inner lagoon margin of the barrier peninsula which forms the inner south shore of the lagoon, the Estero, joins the central sand core of the barrier peninsula of 50–300 m width. This southern inner lagoonal coastline is winding and composed of small, winding channels and sand and mud strips.
Image interpretation: These features are sharply defined (especially on ERTS-MSS-bands 6 and 7).

Environmental features: Opposite the wide sandy core of the barrier peninsula between the beach and the Estero bank, the inner mangrove strip is narrow and not compactly shaped, so that large parts of the shore toward the Estero are built up by barrier beach sand. As a result of the erosion effect of tidal streams in the lagoon, on some parts bluffs 2 m high are formed in front of which a small sandy beach area dries out at low tide. This strip of the barrier island as a landscape unit belongs to the barrier peninsula (terrestrial), as well as to the Estero (inter-tidal) with mangrove vegetation.
Image interpretation: Features are partly recognizable on ERTS-MSS-band 5, the features are partly under minimum visible.

Environmental features: A differentiated feature can be observed near the barrier peninsula's spit point. The barrier peninsula spit on the Estero side is divided by several channels. Accumulation in this area causes a growth of finger shaped ridges up to the High Water level.
Image interpretation: The features on ERTS-MSS-band 5 are only partly distinguishable, partly under minimum visible.

Environmental features: At a distance of 2 km from the spit point, hook-like beach ridges cross the barrier island. About 15 of these landward orientated ridges are lined up from east to west.
Image interpretation: Features are discernable on ERTS-MSS-bands 6 and 7, depending on dry or humid soil.

Environmental features: Wind has caused erosion forms by blowing sand out of the ridges and dunes. Therefore the single walls are separated by cross hollows. Their association can be recognized clearly in the aerial photograph, but only with difficulty in the field (reconnaissance). In other spots in the vicinity the blown-out sand has been compacted into small sand dunes whose height is between 0.5 and 1 m and whose length at the base is between 1 and 3 m.
Image interpretation: Features not identifiable.

Conclusion: Spatially small connections among terrestrial – tidal zones are only detectable when the differences in the substratum such as sand and damp mud give sufficient contrast (as on ERTS-MSS-bands 6 and 7) and are of a size (scale) well over minimum visible.

Geomorphological processes of the terrestrial zone

Environmental features: Comparing ERTS images with aerial photographs of former years one can analyse the development of landforms. The development of the barrier peninsula is very important for changes and development of the coast. Most significant are the beach ridge systems which build up into the barrier island. These beach ridge systems are discernible on the ERTS images. A seaward growth of the coast proceeds by breakers and surf built additional beach ridges. Beach ridges can develop from sandy reefs of surf belt or directly on the beach. A system of up to 20 and 25, and in parts up to 50 successive beach ridges could be determined (by ground truth and aerial photography) on the barrier peninsula San Juan del Gozo.
Image interpretation: Features and pattern of beach ridges are partly identifiable on ERTS-MSS-bands 5, 6, 7.

Environmental features: The core of the island Espiritu Santo is built up of a beach ridge system parallel to the coast stretching from WNW to ESE. The beach ridge system of the islands Espiritu Santo and Madresal formed the old coastline and separated the earliest part of the lagoon from the ocean (about 400 years ago).
Image interpretation: These beach ridge systems can be identified in satellite images: bright grey tone on ERTS-MSS-band 5 in contrast to the mangrove area of the island (dark grey) and as a bright, distinct strip on ERTS-MSS-band 7.

Environmental features: The structure of the island San Sebastian is divided by the channel Ensenada El Artillero into a western and an eastern peninsula. In former times this channel was an inlet between two islands.
Image interpretation: ERTS-MSS-bands 6 and 7: The feature of the interlocking of hook shaped beach ridges of the terrestrial level and the muddy inter-tidal strips and tidal channels on the north side of the eastern part of the island are visible on ERTS-MSS-band 6.

II The inter-tidal zone (between High and Low Water)

Environmental features: Sand areas appear as sand flats in mangrove regions. These lense-shaped sand bodies have tongue-like features and are covered at high water. Fine sand dominates in the sediment composition of these tidal flats, but muddy areas and mixed sediments also appear, so that mangrove vegetation can grow.
Image interpretation: The flats at the data acquisition time of the ERTS satellite image are covered by High Water. On ERTS-MSS-bands 4 and 5 flats can be identified through the transparent water as submarine bottom relief forms, e. g. between the islands Madresal and Pajarito.
Image interpretation: In the tidal flat sediment regions, there are no differences of the substratum identifiable in any of the satellite image bands (spectral bands).

Geomorphological processes in the inter-tidal zone

Environmental features: The small, narrow, sandy island Pajarito is only 10 to 15 m wide on its upper terrestrial body; it is situated on the north slope of the main deep channel of the lagoon, 1 to 2 km south of the island Madresal. Tidal streams with 2 knots velocity pass the southwest beach of the small island Pajarito, so that an erosional process shifts the main channel to its northern slope. Due to sedimentation on the northeastern flat beach of Pajarito island, this island migrates to the northeast.
Image interpretation: On Pajarito island the shallows of the foreshore are almost completely covered by water due to it being High Water Springs at data acquisition time. The island is identifiable on ERTS-MSS-bands 4 and 5 as a white, narrow strip; on ERTS-MSS-bands 6 and 7 there is only a mere hint (limit of minimum visible). A northward shift is evident by comparing the picture of the satellite image of 1973 (ERTS-MSS-bands 4 and 5) with the map of 1954, based on field work of that year, and aerial photograph of 1949 (compare the figure).
Conclusion: This analysis indicates the usefulness of ERTS-MSS images concerning coastal morphology and coastal developments, i. e. long term change detection.

The mangrove forest, economic importance and detection with orbital remote sensing (ERTS-MSS imagery)

The mangrove vegetation covers extended areas of the lagoonal landscape of the Estero de Jiquilisco. Mangrove vegetation grows in the inter-tidal zone. The trunks and crowns of mangrove trees extend in the upper, terrestrial level.

Classification of vertical zones of the Lagoon of Jiquilisco, El Salvador

Depending on High Water and Low Water of prevailing tides with a mean tidal range of 2 m, the environment is separated into vertical zones:

I above High Water level: terrestrial
mainland, barrier islands, and peninsulas with beach ridges and spits and lagoonal islands

II between High Water and Low Water level: inter-tidal
tidal flats and mangrove vegetation and tidal channels; the mangrove vegetation roots in inter-tidal flats but the tops of the trees rise up into the upper level

III below Low Water level: lagoonal and marine
bottom of channels, i. e. submarine topography
and water bodies of the lagoon, i. e. hydrographic features

IV below Low Water level at the entrance and outside of the lagoon: oceanic
offshore shoals, bottom of the sea, i. e. submarine topography and water body of the sea, i. e. hydrographic features

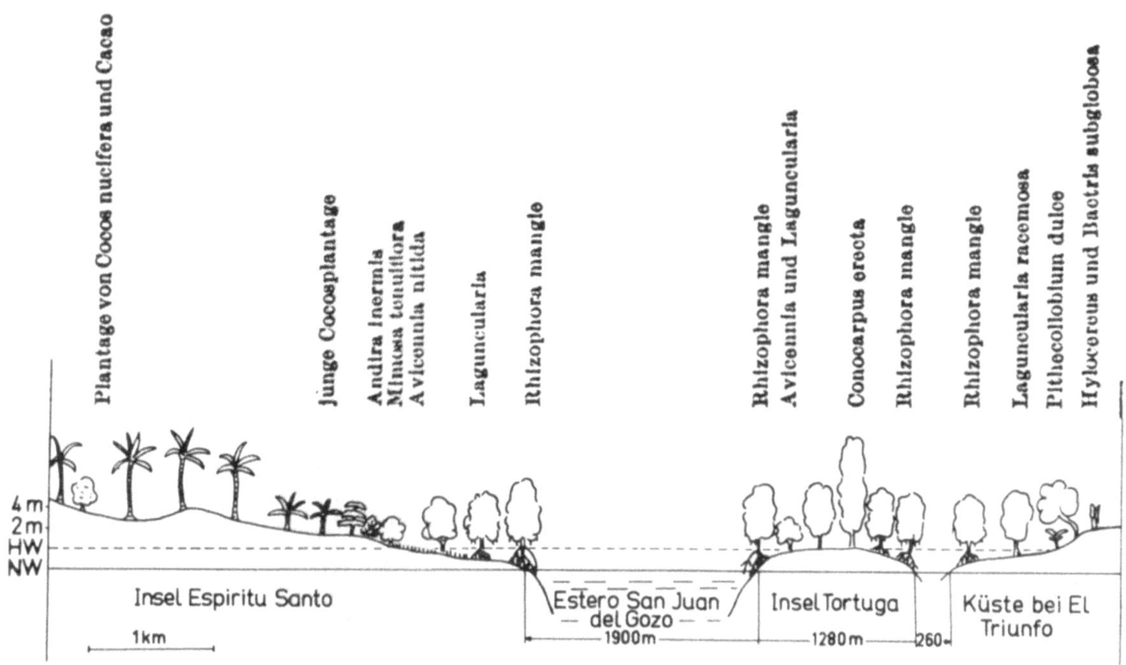

Profile across the main channel of the Estero de Jiquilisco, El Salvador

Zones: terrestrial, inter-tidal (between HW (High Water) and NW (Low Water) and marine mangrove vegetation in the inter-tidal zone
Remark: NW as the German term means "Niedrigwasser" i. e. Low Water

The mangrove stands which compose the only state forests of El Salvador, are used for various purposes. Different species are locally used for housebuilding. It is noteworthy that Rhizophora mangle is used for mining and railroad ties, and that the bark is used for tanning. For further information see: W. Lauer, 1956, p. 44.
Mangrove vegetation can only develop under certain conditions; it is a vegetation form growing in tropical climates in coastal regions. Mangrove vegetation is evergreen the whole year around. This is important for remote sensing detection of vegetation, because there is no optimal period during the year.

The presence of mangrove indicates that certain physical conditions prevail (R. West, 1956). They are:
Tropical temperatures – average temperature of coldest month exceeds 20° C and seasonal range is less than 5° C, daily range less than 5° C, water temperatures always higher than 20° C, and that no arid conditions prevail.
Fine grained alluvium soft mud composed of silt and clay, and rich in organic material.
Mangrove seldom appears on more sandy substratum.
A necessary location precondition against moving water is
a) that tidal water is present – salinity is necessary – (essential precondition) and
b) that there are no breakers (exclusive precondition).

Smaller stands of mangrove even occur in areas of small tidal range. In areas of large tidal range and in arid or seasonally arid climates which are only dampened at spring tides (i. e. sand banks rising high above the average water level), salt may accumulate. As a result the salinity tolerance of mangrove is surpassed, so that these regions remain as islands without vegetation within a mangrove district (Fosberg, 1961). (Compare chapter Golfo de Fonseca of this publication).
A description of the mangrove ecosystems cannot be presented in this paper. Phenologically, bush and shrublike stands are frequent and forest-like stands as in El Salvador seldom develop. The Salvadorean mangrove stands belong to the so-called "western" (Pacific) mangrove type which has fewer varieties than the "eastern" (Atlantic). For species of mangrove vegetation and height see profile. Mangrove vegetation causes the formation of a local climate in its area. Calms and uniform temperatures around 32° C during day and around 25° C at night with humidity between 60 % and 80 % produce a tropical, muggy hot-house atmosphere.

Environmental features: Some units of the detailed vegetation-mapping by W. Lauer, 1956, are identifiable in ERTS images, and some are not (e. g. the salt pans are not detectable). Compare map by W. Lauer which includes a section of the ERTS scene, namely the islands Santo Spirito and Tortuga.
Image interpretation: Individual stands according to the species can only be identified on large scale stereo aerial photographs, but not on small scale ERTS satellite images. The extension of mangrove vegetation was mapped in general from satellite image bands 5 and 6. In ERTS-MSS-band 6 the margin of mangrove vegetation appears best against the upper landscape level (terrestrial, above High Water level), and also in ERTS-MSS-band 6 the margin of the mangrove vegetation against the lower landscape level, water surfaces, is identifiable and can be mapped.
Conclusion: This analysis indicates the usefulness of ERTS-MSS images concerning the mapping of mangrove vegetation.

The water surface of the inter-tidal zone

Environmental features: Tidal streams affect the dynamic and hydrographic features of the entire lagoon. The effect is especially pronounced toward the mouth. The tides determine the upper limit (High Water) and the lower limit (Low Water) of this zone: these correspond to the lower limit of the terrestrial zone, and the upper limit of the marine zone of the landscape, i. e. the inter-tidal zone (horizontal extension) of this level (vertical extension). On the Pacific coast of El Salvador, the tides have a mean spring range of 2 m, and a mean neap range of 1.85 m. The tidal range in the lagoon is modified by local conditions. The tides are of semidiurnal type. The tidal flow is also modified by the lagoon's rather complex bathymetry. Mean sea level for the coast of El Salvador is that of mean sea level at La Union (Golfo de Fonseca).
Image interpretation: Only the inner limit of the horizontal extension of the open water area of this inter-tidal zone can be mapped with the help of bands 6 and 7 of ERTS-MSS image; acquisition time: tidal high water, Mean High Water Springs i. e. tidal inundation. The outer limit could be mapped only at tidal low water. Nevertheless the boundary of the open water area against the mangrove vegetation can be mapped from these ERTS images.

Environmental features: Deep channels and water arms of the lagoon are also flooded at Low Water and as a result form part of the marine zone.
Image interpretation: There submarine bottom topography is not visible on the ERTS images. No differentiation in water depths can be discerned in any of the bands of the satellite image: water areas appear black on ERTS-MSS-bands 6 and 7. Penetration through water occurs in places on ERTS-MSS-bands 4 and 5 where the water is not turbid and opaque, but there is no method of determining from the imagery to what depth the optical visibility penetrates the water transparency. If images of a repetition flight at low water were available, the lower border of the inter-tidal level could be mapped with ERTS-MSS-bands 6 and 7.

Environmental features: The winding course of smaller tidal channels with almost meanderlike pattern is conspicuous. The channels alternate with narrow shoals. This is the typical pattern of the effect of swinging tidal streams. The width of the channels, their water masses and the meander radii are interdependent on each other. Beginning at the last fine branches, the loop radius and canal width increase so regularly, that the connecting link of the outward loop edges from the beginning to the mouth in the main channel is shaped like a trumpet. The bending channels of the innermost parts of the turns fall dry at Low Water. These are hydrographic forms in this inter-tidal level which type is distributed world wide, as e.g. in the lagoons off the Atlantic coast of the USA and France. (Verger, 1968).
Image interpretation: A map of the water area of this level can be compiled at high water especially well with the help of ERTS-MSS-band 7 (infrared), but the smallest branches of tidal channels are not visible on the ERTS image, so that their network compiled with ERTS images is generalised in comparison with maps of the same scale, but based on large scale aerial photographs.

The water bodies

The complex conditions of two-medium photography and muddiness and colouration of water bodies and seawater cannot be discussed here. Seawater represents a mixture of solution, colloid, suspension and content of biological substratum and living creatures. Specific lighting conditions also influence the transparency of seawater.

Environmental features: The brown-yellow colour (in nature) of the lagoonal water is caused by its load of suspended matter, which makes the water so turbid that it is very opaque (during the rainy season the creeks coming from the coastal plain pour fresh water into the Estero, so that density levels result from the differing salinity). The water masses of the Estero are moved and mixed twice daily with the tides: the lagoon "breathes" large amounts of water in and out twice daily, as a pumping system of lagoonal and sea water interface.
Image interpretation: Differentiation of water bodies in the lagoon cannot be detected in any of the bands of ERTS-MSS images.

III The marine zone lagoonal and oceanic (below Low Water)

Nearshore dynamics as investigations of tidal inlets and sediment interaction on beaches, estuaries and shelves and mesoscale studies of shelf processes, i. e. long term change detection.

The breaker and surf zone

The morphologically most important effects on the coast are caused by the breakers and surf and its secondary phenomena, the surf currents.
Surf is the wave activity in the area between the shoreline and the outermost limit of breakers.

Environmental features: Breakers on the coast of El Salvador reach maximum heights of 3 m, whereas the mean values are 1 to 1.5 m. According to the interpretation of aerial photographs surf waves with heights of 1.50–1.80 m which break uniformly in a 200–1000 m extension occur there very often. The angle between wave fronts and coast is usually not larger than 10°. Depending on the slope of the beach spilling breakers, plunging breakers and surging breakers were observed during ground observations on this coast.
Image interpretation: Breakers of the surf zone are identifiable (in front of the coast of the barrier beach of the peninsula San Juan del Gozo, in the middle of the image section). It shows best on ERTS-MSS-band 4, not so well on ERTS-MSS-band 5. It is worthwhile noting that the surf zone shows up well on ERTS-MSS-bands 4 and 5 due to its white foam caps, but does not appear on ERTS-MSS-bands 6 and 7 (infrared) because of the remission and reflection properties of the strongly absorbing substratum water concerning this spectral band of MSS-channels.

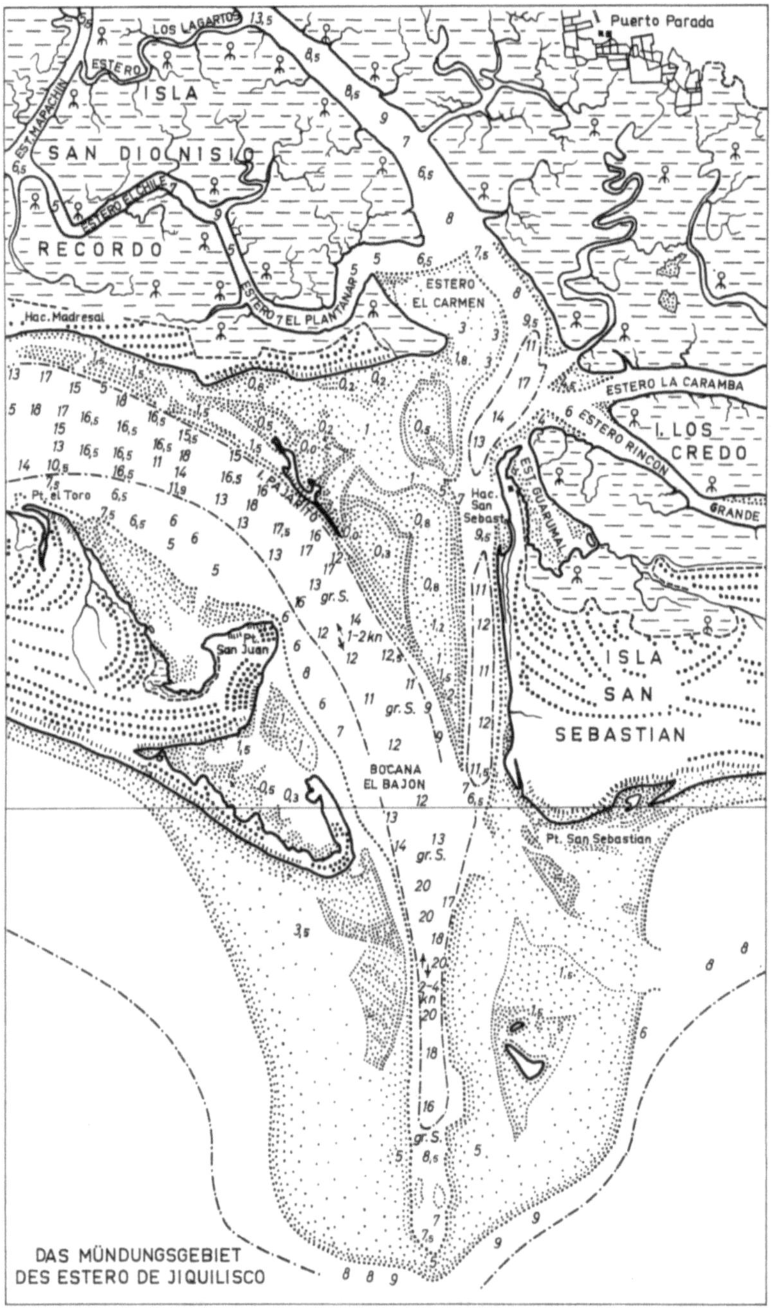

Entrance and approaches of the Jiquilisco lagoon, Bocana el Bajon

scale 1 : 80 000

main channel, offshore and foreshore shoals, sandbanks, islands, beach ridges on barrier islands and mangrove vegetation

as mapped and drawn by H. G. Gierloff-Emden, 1955

Aerial photograph of the Lagoon of Jiquilisco, El Salvador, with the Bocana el Bajon
scale of the aerial photograph, taken in 1949, is 1 : 20 000
scale of this printed picture is 1 : 50 000
regions of the entrance, presented for detail of features and patterns and comparison with ERTS–1 MSS
image mosaic of aerial photographs of Dirección de Cartografía, San Salvador

Morphological processes in the marine zone

Example: Entrance and shoals of the lagoon: Bocana el Bajon (compare map).

Environmental features: The 2 km wide mouth of the Estero de Jiquilisco is formed by the Bocana el Bajon. When leaving the mouth of the Estero the ebb-tide stream due to the jet effect, is concentrated into a stream which, gyroscopically stabilized, extends 4 km seaward. Current speeds of up to 4 knots are then reached. This explains the 20 m deep, 300 to 400 m wide channel mouth. The phenomenon of deviation of the axis of the channel compared to the axis of the inner-lagoonal channel and pattern of marginal channels is generally described by Samojlov, 1956.

The bar in front of the mouth: The ebb current has built up a delta-like shoal pattern to seaward. In 1954 there was a 5 km long and 4 km wide sand bar near the mouth, depths were less than 5 m, and in several places less than 1 m. On nautical charts and in sailing directions the area is called "Barra de Lempa" or "Bajos Lempa".

Image interpretation: From 1954 to 1973 further changes of the Bocana and the bar took place as it is documented with the ERTS satellite images in ERTS-MSS-bands 4 and 5. According to these images the exit of the bar presently lies exactly south of the west corner of San Sebastian island, i. e. it has migrated or shifted to the east. The bar is well visible due to the surf (white strip on ERTS-MSS-bands 4 and 5, not on ERTS-MSS-bands 6 and 7).

Environmental features: On both sides of the entrance to the mouth sandy hooks and long tongue-like sand banks form due to sedimentation.

Image interpretation: Discernible by penetration through the water on ERTS-MSS-bands 4 and 5.

Environmental features: The 10 and 20 m depth contours off the coast show a seaward shift caused by the delta at the mouth of the Estero de Jiquilisco (compare map).

In 1952 there was a 500 m long island partly covered with mangrove, 2,5 km south of Punta San Sebastian (in position 13° 8′ 30″ N and 88° 26′ 30″ W) (compare map).

Image interpretation: This island, mapped during field work in 1954, has been shifted and decomposed.
A sandbank, still a "bar", is identifiable only on ERTS-MSS-bands 4 and 5 (penetration) not on ERTS-MSS-bands 6 and 7 as it is covered by water.

Sea swells and wind waves

With great frequency and regularity, distinct pronounced swell waves have been observed on the coast of El Salvador with usually low wave height, wave length of 80 m to 120 m, and long periods, more than 10 sec.

Image interpretation: Swell is not discernible on any of the bands of the satellite image. For comparison, the aerial photograph from 1954 can be used. On this, surf and wind waves are perceptible.

Water bodies and rip currents in the ocean

Rip currents: The occurrence of rip currents (undertow) from the beach 300 m seawards could be observed on aerial photographs on this coast.

Image interpretation: Rip currents are not visible on the ERTS image, even though they are surely present and according to their size, lie above minimum visible. The contrast is obviously insufficient; they ought to be visible on ERTS-MSS-band 4. Rip currents on the coast of El Salvador are visible in the region of the Golfo de Fonseca on ERTS-MSS images.

Visibility of suspended sediments

Water masses, plumes and sediment transport discriminated by remote sensing techniques.
Water masses in the sea in front of the coast are only visible on ERTS-MSS-band 4 of the satellite image. They reach, visible as grey clouds, up to 8 km seawards off the coast. The turbid water masses do not present a prominent plume-like pattern.

Environmental features (ocean): To seaward of the lagoon the 10 m depth contour lies about two or three km, and the 20 m depth contour about five to six km, offshore. Both depth contours show seaward shifts in front of the mouth of the lagoon (Bocana). The outer limit of the water mass therefore lies far offshore.

Image interpretation: In the ocean area directly to seaward of the coast of El Salvador there are no constant ocean currents. The distance offshore of the limit of the muddy water mass off the coast is 8 km. The large size of this muddy water mass points to a different connection: when the lagoon water discharges with the ebb, one knot tidal stream, a distance of 8–10 km to seaward per tide (6 hours) is reached. The muddy water mass from the previous ebb stream drifts along the coast. It will probably not be material picked up by breakers, because the fraction of the sands on the beach and foreshore are too large to remain in suspension for such a long distance (substratum is usually moved along the ground), though such a possibility cannot be excluded until an analysis of water samples has been made.
Therefore it seems to be proved that the grey strips on ERTS-MSS-band 4 are not shapes of submarine topography, but are muddy water bodies with sediment clouds.

NASA ERTS-1-MSS imagery colour composite

from multispectral scanner images
bands 4, 5, 7, enlarged section of NASA ERTS-1-MSS 15/1088 ID 1209-15444

Jiquilisco Lagoon of El Salvador

(Vegetation comes out in red colour)
(Colour composite made by Department of Geology, University of Munich)

Producing small scale thematic maps from ERTS-MSS imagery: application of the ERTS imagery
Map of mangrove vegetation and hydrographic map

ERTS–MSS–C.:4

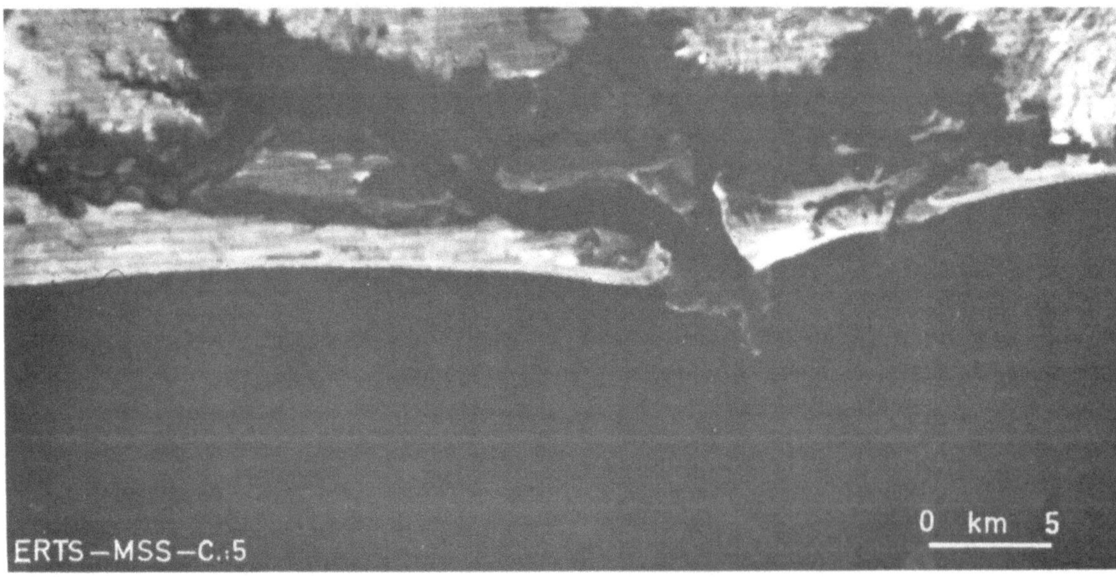
0 km 5
ERTS–MSS–C.:5

km
0 1 2 3 4 5
Mangrovegebiete

Thematic map showing the mangrove vegetation of the inter-tidal zone in the Jiquilisco lagoon, El Salvador

H. G. Gierloff-Emden, 1973
Scale approximately 1 : 300 000

Produced from enlarged section of NASA–ERTS–1 multispectral scanner image

by technic of grey tone scale extraction (in coincidence with the thematic extraction)
MSS-200 15/1088, observation ID 1209-15444, date of acquisition: 17. Feb. 1973
Tidal condition: High Water

MSS-imagery used:
bands 6 and 7 due to high contrast between water surface and terrain including inter-tidal mangrove vegetation,
bands 4 and 5 due to the differing representation of mangrove vegetation and the terrestrial zone vegetation

Thematic presentation of cartography: mangrove vegetation map

dark grey: mangrove vegetation of the inter-tidal zone in which the crown cover projects above the level of the inter-tidal flats but which is considered a feature of the inter-tidal zone,
light grey: terrestrial zone (land including mainland and barrier peninsulas, barrier islands and spits),
white: open water surfaces of the Pacific Ocean, the lagoons and waterfilled channels within the mangrove vegetation zone including smaller branches of the channels of the water channel and inter-tidal water-filled channel system down to 30–60 m in width
Compare the details of the mangrove vegetation of this thematic map made from ERTS-MSS imagery with the details of the "Map of Estero de Jiquilisco", some pages before

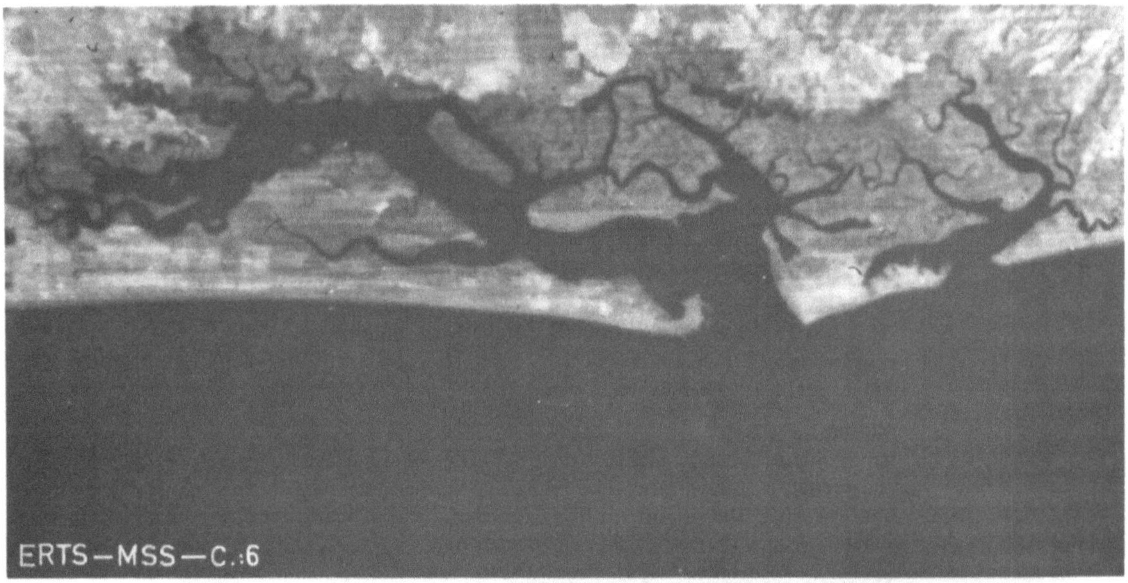
ERTS—MSS—C.:6

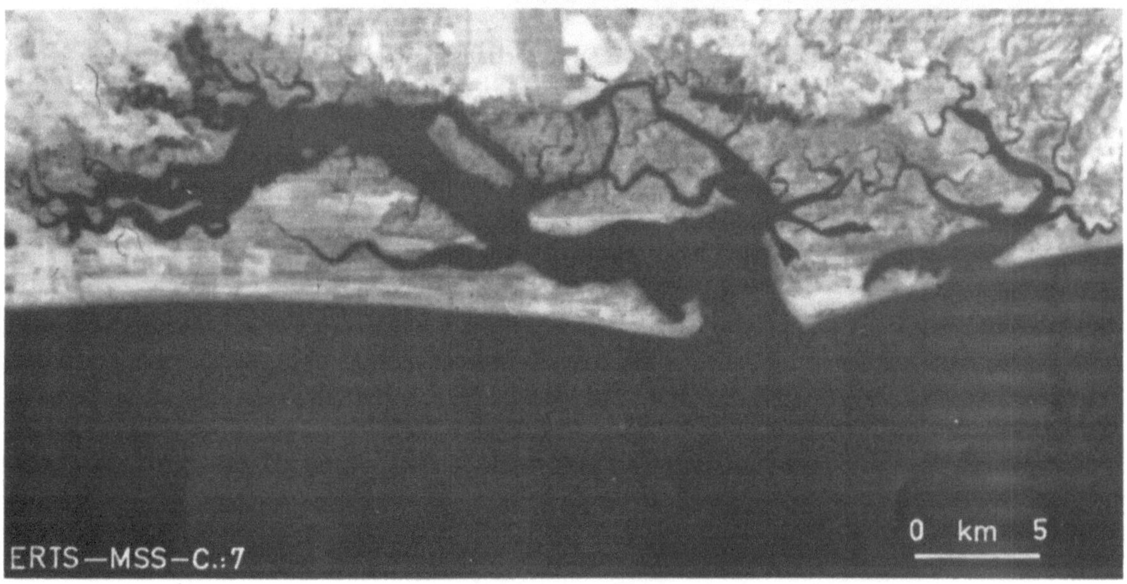
0 km 5
ERTS—MSS—C.:7

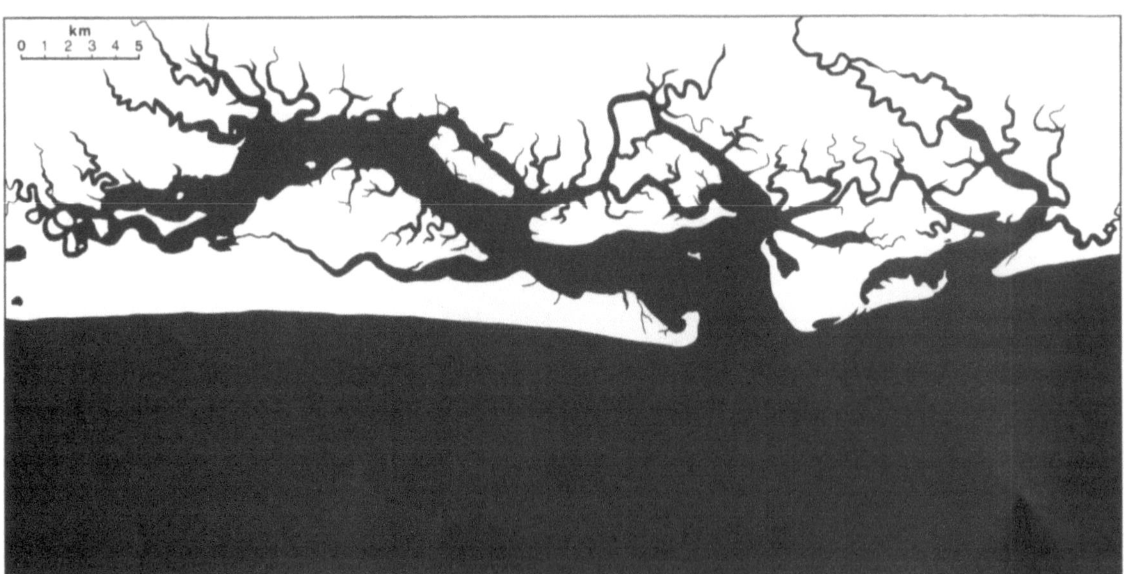
km
0 1 2 3 4 5

Thematic map of the water surfaces of the Jiquilisco lagoon, El Salvador

H. G. Gierloff-Emden, 1973

Scale approximately 1 : 300 000

Produced from enlarged section of NASA–ERTS–1 multispectral scanner image

by technic of grey tone scale extraction (in coincidence with the thematic extraction)

MSS-200 15/1088, observation ID 1209-15444, date of acquisition: 17. Feb. 1973
Tidal condition: High Water

MSS-imagery used:
band 7 (due to high contrast between water surfaces and terrestrial terrain) and band 6 (due to high contrast between water surface and inter-tidal mangrove vegetation)

Thematic presentation of cartography: hydrographic map

black: open water surface of the Pacific Ocean, the lagoons, and waterfilled inter-tidal channels within the mangrove vegetation zone, including smaller branches of the water channel system down to 30–60 m in width; smaller dendritic water channels and ditches are minor as the minimum visible on the ERTS image

white: terrestrial terrain and mangrove vegetation of the inter-tidal zone (in which the crown cover projects above the level of the inter-tidal flats but which are considered as features of the inter-tidal zone)

Compare the details of the tidal channels of this thematic map made from ERTS-MSS imagery with the details of the "Map of Estero de Jiquilisco", some pages before

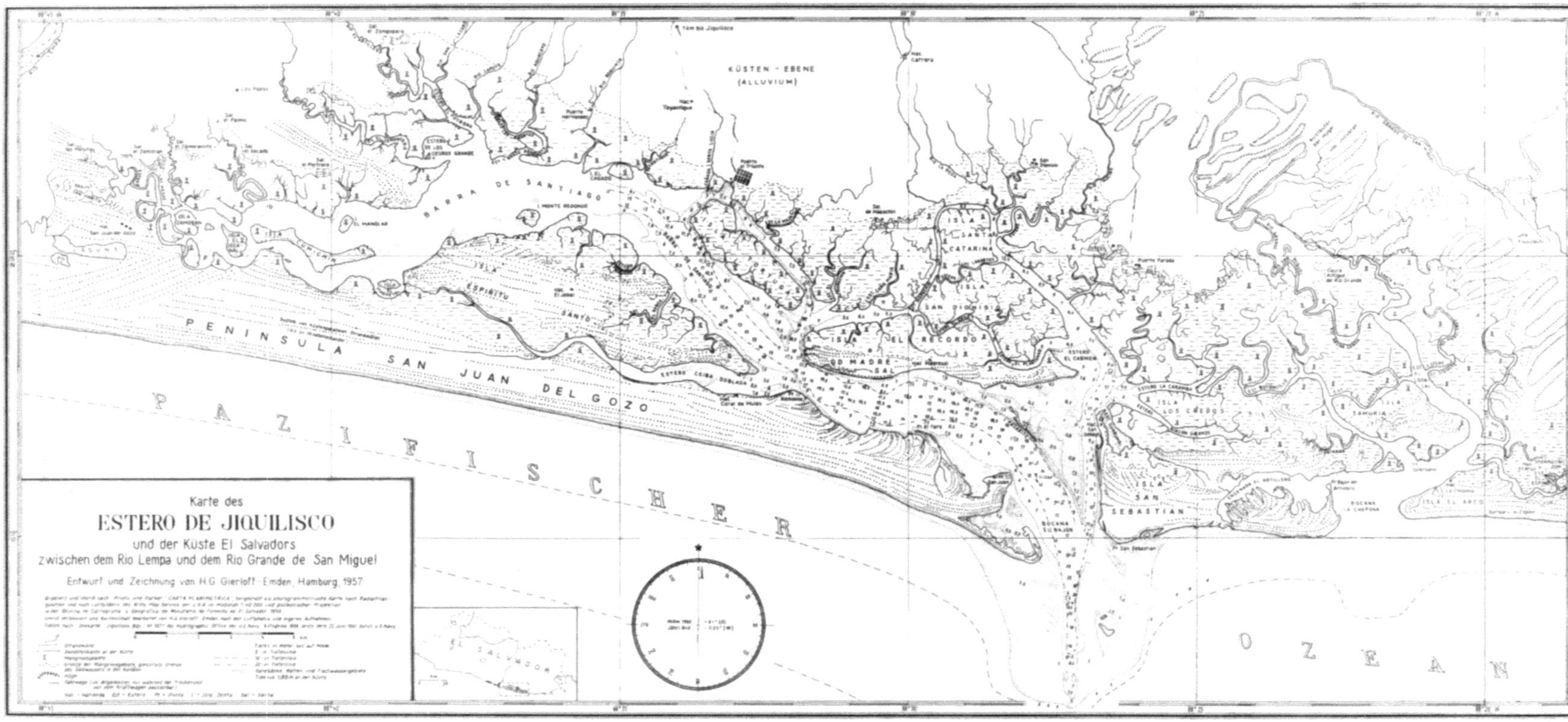

Map of Estero de Jiquilisco, Coast of El Salvador

Scale 1 : 210 000
compiled and drawn by H. G. Gierloff-Emden after fieldwork and aerial photograph
presenting: barrier peninsula and barrier islands, with beach ridges (dotted lines), mangrove vegetation and lagoonal water body, water tidal channels

Compare the details of the tidal channels and of the mangrove vegetation of this map with the small scale thematic maps made from ERTS-MSS imagery on pages before.

Producing small scale thematic maps from ERTS-MSS imagery: application of the ERTS imagery
Map of mangrove vegetation and hydrographic map

On the basis of interpretation and thematic extraction of the use of images of 4 bands of ERTS-1 multispectral scanner (MSS) imagery, thematic maps on a scale of approximately 1 : 300 000 presenting water areas and mangrove vegetation of the lagoon "Estero de Jiquilisco" were compiled. Ground truth is based on pre-information: intensive field work was carried out many years before. H. G. Gierloff-Emden, 1959. The production of a map of a scale of 1 : 300 000 means that this is 10 times larger than the original MSS imagery scale. For the mapping of water and mangrove areas of a large lagoon on this scale, the ERTS-MSS satellite imagery is considered to be quite suitable concerning the accuracy and details, and it is far superior to methods of classical surveying concerning the time of production.
However, for lagoons of less than 5 km length and 2 km width, this orbital remote sensing material is not appropriate. A comparison of the ground-checked map, prepared from large scale aerial photographs, with the new maps from the ERTS images was carried out. It has been concluded that the ERTS multispectral scanner imagery is quite suitable for the given task. This task was the preparation of small scale maps of the coastal region on which the boundaries of tidal landscapes, hydrological and vegetation classes have been indicated in order to allow the surface area for each to be estimated.
The small tidal channels and dendritic branches of the tidal and inter-tidal waterways in the mangrove landscape were also studied. The representation of such 40–60 m wide linear features on a map of scale 1 : 300 000 amounts to about 0,2 mm. Due to their linear form, such tidal channels, although in exceptional cases having a width as small as 30 m may also be identifiable on ERTS-MSS satellite images. At this point however, the resolution limit of the ERTS-MSS imagery has been reached, including also ERTS-MSS-band 7 with strong contrasts. Smaller tidal channels cannot be mapped from this imagery. Therefore this is effectively the limit of information content for landscape features on such satellite images and the limit for automatic extraction of such features from a satellite image.
On the 1 : 300 000 scale map the small tidal channels can only be represented by exaggerated symbols, since 0,2 mm is also the limit of drafting accuracy. This would imply suppression of details for the purpose of representing the character of the landscape. In the present case, this character relates to the tidal flat and mangrove lagoon landscape. It is expressed in nature as numerous fine dendritic branches and ditches of the tide-filled channel system. An analogous example is the representation of drainage ditches in the polder landscape on the topographic maps of same scale. Compare these thematic maps with the "Map of Estero de Jiquilisco" (opposite page).
Application: Detailed expression may be important for water transportation purposes, for example, in connection with fishing and waterway transport. Such maps of waterways cannot be prepared from ERTS-MSS images if the smallest waterways are required. Referring to the build up and development of such tidal channels, the satellite image can well serve as a check on the changes in the larger waterways. However, concerning the estimation of water surface areas (open areas and larger channels) of large lagoons, the ERTS-MSS images are a sufficient source of data. This also applies to the estimation of water covered area and of mangrove vegetation, with which the tidal channel system is closely connected.
In the case of the seaward limit of this landscape the beach has been drawn as a distinct boundary line representing the shoreline of the coast. This is based on the sharp land – water contrast available in the imagery from ERTS-MSS-bands 6 and especially 7, and has a accuracy of about 60 m.
With the help of the ERTS photographs it was possible to investigate long term morphological processes on the coast, useful for prediction of nearshore coastal environments. For example, using ERTS-MSS-bands 4 and 5, the modifications of outlets and sandbars associated with the lagoons could be studied.
Topography of the microscale, i. e. ripples or cusped spits of the size of some metres only, cannot be detected with this ERTS-MSS images, but also could not be drafted in scale on a map of a scale of 1 : 300 000. Such maps could be gained by thematic extraction by photographic separation of grey tone, i. e. *automatic* mapping by photographic processing to save time and compilation work.
Therefore the application of ERTS-MSS imagery is very useful and economic for thematic extraction with the task of the production of thematic maps in large scale of lagoons with their waterways, coastlines and vegetation.

Mapping the land-sea boundary for large scale and small scale charts

Charting the seaward limit of this coastal environment, the beach has been drawn as a distinct boundary line, representing the shoreline of the coast.

This method may be used for small scale charts mapped with orbital remote sensing imagery of the type of ERTS–1, especially where the coast is still imperfectly known or shoreline unsurveyed.

For large scale maps, the boundaries have to be mapped according to standard terms of Admiralty Charts and of coastal geomorphology of the small scale features; also precise tidal data are required. These data are still not available however for much of the earth's coastline.

Mapping obscured tidal boundaries

"Demarcating and mapping boundaries formed by, or dependent upon, tidal datum planes present persistent and difficult problems, particularly in areas where the boundaries are obscured by some type of vegetation.

While both the Mean High-Water and Mean Low-Water lines play important roles in coastal boundary determinations, efforts are being focused on the Mean High-Water line due to its importance in the control of wetlands and in coastal zone management. Only a relatively small part of the total length of the Mean Low-Water line is obscured by vegetation, making it a minor problem in comparison to the Mean High-Water line.

The mean high-water line in dense mangrove and in coastal marsh is almost invariably obscured by indigenous vegetation, especially in data acquired with remote sensors. Vegetation that obscures the boundary combines with the characteristics of its environment to make terrestrial operations extremely difficult, if not totally impractical. Together, these factors create problems for the cartographer, engineer, and surveyor".
Huebner, G. L., 1975, p. 1603, "The Marine Environment", Chapter 20 in Manual of Remote Sensing.

Bibliography

(compare also general bibliography)

Gierloff-Emden, H. G., 1959, Die Küste von El Salvador, eine morphologisch-ozeanographische Monographie: *Acta Humboldtiana, Series Geographica et Ethnographica*, vol. 2, Wiesbaden, 183 p., 76 fig., 14 maps (extended bibliography).

Gierloff-Emden, H. G., 1960, Küstenmorphologie mit Hilfe des Luftbildes: in: Das Luftbild in seiner landschaftlichen Aussage, vol. 3 of *Landeskundl. Luftbildausw. im mitteleurop. Raum*, Bad Godesberg.

Gierloff-Emden, H. G., 1961, Nehrungen und Lagunen, Gesetzmäßigkeiten ihrer Formenbildung und Verbreitung: *Pet. Mitt.*, 105. Year of Publ., vol. 2 and 3, Gotha.

Gierloff-Emden, H. G., 1974: Anwendung von Multispektralaufnahmen des ERTS-Satelliten zur kleinmaßstäbigen Kartierung der Stockwerke amphibischer Küstenräume am Beispiel der Küste von El Salvador. *Kartographische Nachrichten*, vol. 24, no. 2, p. 54–76, Bad Godesberg.

Koeman, C., 1970, The impact of photography from space on small scale and atlas cartography: in Intern. Jahrb. f. Kartographie, p. 35–40.

Latham, J. P., 1973, A comparison of interpretation and photogrammetric methods for delimiting the mean High Water position on a tropical beach: *Proc. Am. Soc. Photogr.*, Symp. on Remote Sensing in Oceanogr., Washington D. C., p. 803–815.

Lauer, W., 1956, Vegetation, Landnutzung und Agrarpotential in El Salvador: *Schrift. Geogr. Inst. d. Univers. Kiel*, vol. XVI, no. 1, 98 p.

Lewis, A. J. and H. C. Macdonald, 1972, Mapping of mangrove and perpendicular-orientated shell reefs in southeastern Panama with side-looking radar: *Photogrammetria*, vol. 28, p. 187–199.

Reimold, R. J., J. L. Gallagher and D. E. Thompson, 1973, Remote sensing of tidal marsh: *Photogramm. Engineer., Am. Soc. Photogramm.*, vol. 39, no. 5, p. 477–489.

Samojlov, 1956, Die Flußmündungen: Gotha, 647 p.

Verger, F., 1968, Marais et wadden du littoral français, étude de géomorphologie, Bordeaux, 541 p.

Wobber, F. J. and R. R. Anderson, 1972, Operational wetlands mapping using multiband aerial photography: Coastal Mapping Sympos., *Am. Soc. Photogramm.*

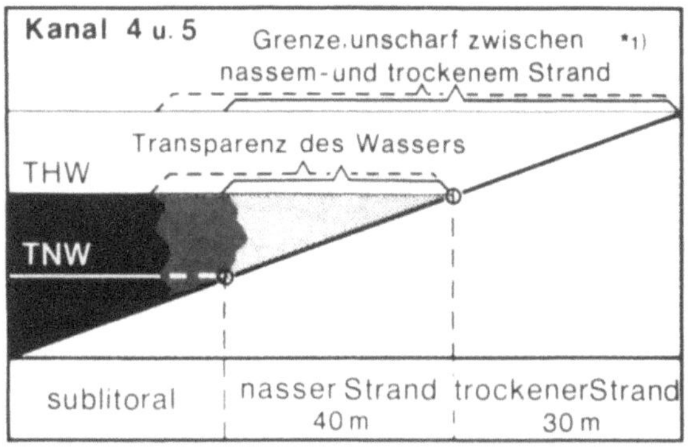

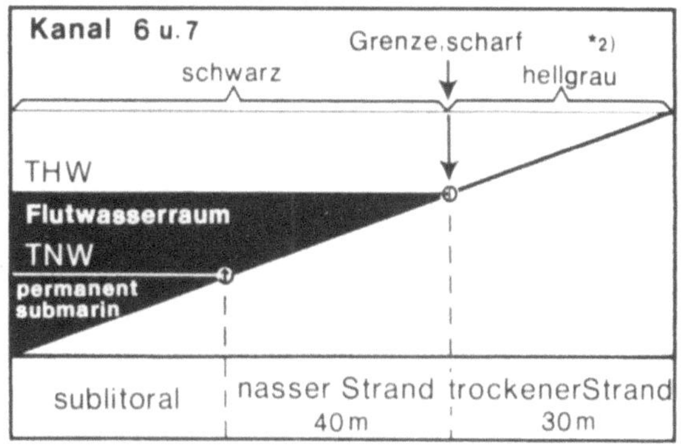

Translation of the German terms of the above Figures:

Figure:	Figure:
Band 4 and 5 borderline not clearly defined between wet and dry beach	Band 6 and 7 borderline sharp black light grey
Transparency of the water THW = Tidal High Water TNW = Tidal Low Water	 THW = Tidal High Water inter-tidal water volume TNW = Tidal Low Water permanent submarine
sublittoral foreshore shore wet beach dry beach 40 m 30 m	sublittoral foreshore shore wet beach dry beach 40 m 30 m

Problems of mapping the coastline (land – water boundary on the tidal coast) using orbital remote sensing systems, NASA–ERTS–1 MSS imagery

(advantage of the 4 different bands)

H. G. Gierloff-Emden 1973

Coastal boundary surveys

Due to the transparency of the water on bands 4 and 5 (green and orange spectral band) light sand substrate of shore and foreshore instead of a sharp bordered waterline is reflected:

1. dry sand beach (shore) above THW (Tidal High Water)
2. wet sand beach (foreshore) THW to TNW (Tidal Low Water), i. e. inter-tidal zone
3. oceanic bottom as sublittoral below TNW (Tidal Low Water) according to transparency

This means a wider zone composed of elements of the tidal margin and the oceanic bottom is reflected on bands 4 and 5. The latter appears as a light border on the picture above to minimum visibility but reflecting not only the beach but also a submarine belt, i. e. not the exact land – sea boundary.

Water occurs almost pure black on the infrared multispectral scanner MSS band 7, whereas the beach occurs in light grey or white. Due to this effect the separation of land and sea is depicted with high contrast and can be used especially for coastal boundary survey. This is important, since coastal zones of the earth are being given new emphasis caused by the importance of exploitation of the benefits of coastal zones such as mineral resources and fisheries.

Due to the high absorption of water on bands 6 and especially 7 (red and infrared), there is a sharp contrast on images of the bands between water and land. This means: the possibility of mapping a sharp limit and clear boundary of the dry sand beach (shore) on a very small zone which, because of its small extent and linear nature shows a sharp contoured boundary between continental and oceanic features.

Bibliography of the manual

Handbooks

NASA, 1967, Earth photographs from Gemini III, IV and V, 245 photographs in colour: NASA Spec. Publ. 129, Scientific and technological informations Dir. NASA, Washington.

NASA, 1969, Earth photographs from Gemini VI through XII: NASA Spec. Publ. 171, Washington 1969.

Colwell, R. N., (ed.), 1960, Manual of photographic interpretation, Washington, D. C. American Soc. Photogramm., 868 p.

Manual of remote sensing, 1976, ed. by R. G. Reeves, 2 vol.: Am. Soc. Photogramm., Falls Church, Va.

Smith, J. T. (ed.), 1968, Manual of colour aerial photography, cap. IX Colour photography from space, p. 365: Am. Soc. Photogramm.

Thompson, M. M. (ed.), 1966–67, Manual of photogrammetry, 3. ed., vol. I, 536 p., vol. II 1199 p.: Am. Soc. Photogramm.

UNESCO, 1968, Aerial surveys and integrated studies: Proceedings of the Toulouse conference, United Nations Educational, Scientific and Cultural Organization, Paris.

The great project, space exploration and Apollo program: scientific insights and tangible benefits derived for mankind, 1971, Carl Zeiss, Oberkochen (ed.), Weinbrenner Verlag, Stuttgart.

General information on remote sensing

Barret, E. C., 1974, Climatology from satellites: Methuen & Co. Ltd., London, 418 p.

Fischer, W. A., 1969, Results of earth resources surveys of the earth and plans for future satellites: presented to joint meeting of the American Astronautical Society and the Operations Research Society, June 1969, reprint.

Gierloff-Emden, H. G. and J. Bodechtel, 1969, Weltraumbilder der Erde: Paul List Verlag, München, 176 p.

Gierloff-Emden, H. G. and J. Bodechtel, 1974, Weltraumbilder, die dritte Entdeckung der Erde: Paul List Verlag, München, 207 p.

Gierloff-Emden, H. G. and H. Schroeder-Lanz, 1970–71, Luftbildauswertung, 3 vol.: B. I. Hochschultaschenbücher, no. 358/358a, 367/367a, 368a/b, Bibliographisches Institut, Mannheim, vol. I 154 p., vol. II 303 p., vol. III 499 p.

Gregory, A. F., 1971, Earth observation satellites, a potential impetus for economic and social development: UNO, *World Cartography*, vol. XI, p. 1–15, New York.

Hempenius, S. A., 1976, Critical review of the status of remote sensing: *B. u. L.*, vol. 1, p. 29–42.

Kondratjev, K. Y., E. P. Borisenkov and A. A. Morozkin, 1970, Interpretation of observation data from meteorological satellites: Translated from Russian by A. Wald and M. Levi, ed. by P. Greenberg, Israel Program for Scientific Translations, Jerusalem, 370 p.

Konecny, G., 1975, Approach and status of geometric restitution for remote sensing imagery: *B. u. L.*, vol. 1.

Lowman, P. D., , Space panorama.

Schneider, S., 1974, Luftbild und Luftbildinterpretation: in Lehrbuch der Allgemeinen Geographie, vol. XI, Walter de Gruyter, Berlin–New York, 530 p.

Verstappen, H. T., 1970, Orbital photography and its uses: *Geoforum*, 1, 2.

Verstappen, H. T. and R. A. van Zuidam, 1968, ITC-system of geomorphological survey: *ITC-textbook of photointerpretation*, Vol. VII, 2, 49 p.

Winogradow, B. W., 1966, Weltraumphotographie für die geographische Erforschung der Erde: in *Nachr. Geogr. Ges.*, vol. 98, 2, p. 101–111 (Russ.)

Wittbrodt, H., H. Mielke, G. Narimanow and J. Saizew, 1975, Raumfahrt für die Erde: Transpress VEB, Berlin, 320 p.

Wyszecki, G., 1960, Farbsysteme: Musterschmidt Verlag, Göttingen, 144 p., 56 fig.

Remote sensing concerning oceanography and coasts

Badgley, P. C., L. Miloy and L. Childs, 1969, Oceans from space: Gulf Publ. Comp., Houston, Texas, 234 p.

Ewing, G. C. (ed.), 1965, Oceanography from space: The Woods Hole Oceanogr. Inst., Woods Hole, Mass.

Gierloff-Emden, H. G., 1960, Küstenmorphologie mit Hilfe des Luftbildes: in Das Luftbild in seiner landschaftlichen Aussage, *Landeskdl. Luftbildausw.*, vol. 3, p. 29–32.

Gierloff-Emden, H. G., 1971, Auswertung von Satellitenbildern zur Meereskunde: in The great project, Carl Zeiss (ed.), Weinbrenner Verlag, Stuttgart.

Jerlov, N. G., 1968, Optical oceanography: Elsevier Publ. Comp. Amsterdam.

Jerlov, N. G. and E. Steemann Nielsen (ed.), 1974, Optical aspects of oceanography: Academic Press, London–New York, 500 p.

Klemas, V., D. Bartlett and R. Rogers, 1975, Coastal zone classification from satellite imagery: *Photogramm. Engineer. Remote Sensing,* vol. 41, p. 499-507, 509-513.

Mairs, R. L., 1970, Oceanographic interpretation of Apollo photographs: *Photogramm. Engineer.*, vol. 10, p. 1045.

Orme, A. R. and L. L. Loener, 1974, Remote sensing of subtropical coastal environments: Natal, South Africa, Off. Naval Res. Techn. Rep. no. 3, 97 p.

Stevenson, R. E. and R. M. Nelson, 1968, An index of ocean features photographed from Gemini spacecraft: Bur. Commer. Fish. Biol. Lab., Contr. no. 235, 348 p.

Szekielda, K. H., 1970, The development of upwelling along the Somali coast as detected with the Nimbus 2 and 3 satellites: Goddard Space Flight Center, Greenbelt, X-651-70-410, 52 p.

Thomson, F. J., 1974, Basic investigations for remote sensing of coastal areas: Quarterly report, 15. July–15. October 1974, Environmental Res. Inst. Michigan, Ann. Arbor.

Verger, F. et al., 1973, Les méthodes de la télédétection appliquées au littoral français: Séance du 1er décembre 1973, *Bull. Assoc. Géogr. Franç. Paris,* no. 411–412, p. 695–729.

Verger, F. et al., 1973, Marshes and turbid waters in the French Atlantic littoral: NASA C. R. 130800, Febr. 28, 43 p.

Wieneke, F. u. G. Kritikos 1971: Strandnaher Wasser- und Materialtransport. *Umschau*, no. 24, p. 169–170.

Yost, E. and S., 1970, Remote sensing of coastal waters using multispectral photographic techniques: Long Island University, 209 p.

Final Report on the space – oceanographic study, Dec. 1967 and March 1968, prepared for the National Council on Marine Resources and Engineering Development by the General Electric Co., Missile and Space Div., Valley Forge Space Center, Philadelphia.

ERTS imagery

NASA earth resources technology satellite, document 71504249, data users handbook, Goddard Space Flight Center, ed. 15. Sept. 1971 and revisions from 15. Feb., 4. May, 18. July, 8. Sept., 15. Sept. and 17. Nov. 1972.

Earth resources technology satellite: Ref. Manual, General Electric Space Div., Philadelphia.

Cf. die symposia, Third annual earth resources program review, vol. I, II, III, 1970: NASA Manned Space Craft Center, Houston, Texas, and Fourth annual earth resources program review, vol. I, II, III, IV, V, 1972: NASA Manned Space Craft Center, Houston Texas.

Earth resources technology satellite – 1. symposium proceedings, Sept. 29, 1972, Goddard Space Flight Center, 1973, 165 p.

NASA earth resources survey program, National Technical Information Service, Weekly government abstracts.

Colvocoresses, A. P., 1970, ERTS-A satellite imagery: in *Photogramm. Engineer.* vol. 6.

Ewing, G. C., 1973, A preliminary assessment of ERTS imagery for marine resources: NASA symposium, March 5–9 1973, vol. II.

Keller, D. W., 1970, Earth resources satellite systems: in *Journ. Brit. Interplan. Soc.*, vol. 4, p. 257–274.

Klemas, V., D. Bartlett and R. Rogers, 1974, Skylab and ERTS–1 investigations of coastal land use and water properties: AIAA/AGU conference on scientific experiments of Skylab, Huntsville, Alabama, Oct. 30 – Nov. 1, 1974.

Short, N. M. and N. H. MacLeod, 1972, Analysis of multispectral images simulating ERTS observations: NASA X-430-72-118, Goddard Space Flight Center, Greenbelt, Maryland.

Coasts, general

Bird, E. C. F., 1964, Coastal landforms, an introduction to coastal geomorphology: The Australian Nat. Univers., Canberra.

Chapman, V. J., 1960, Salt marshes and salt deserts of the world: Interscience Publish., New York, 392 p.

Dietrich, G., 1939, Das amerikanische Mittelmeer, ein meereskundlicher Überblick; *Zeitschr. Erdkde*, vol. 3–4, p. 103–130, map, fig.

Dietrich, G. and K. Kalle, 1965, Allgemeine Meereskunde: Borntraeger Verlag, Berlin (2. ed.).

Fairbridge, R. W. (ed.), 1966, The encyclopedia of oceanography, encyclopedia of earth sciences series, vol. I: Reinhold Publish. Corp., New York.

Gary, M., R. McAfee and C. L. Wolf (eds.), 1972, Glossary of Geology: Am. Geol. Inst., Washington D. C., 857 p.

Guilcher, A., 1954, Morphologie littorale et sous-marine: Presse Universitaires de France, 247 p.

Johnson, D. V., 1919, Shore processes and shoreline development: Wiley & Sons, New York.

Kerr, A. J. and Kordick, A., 1972, Oceanic Cartography, *ICA Public.*, Amsterdam, 131 p.

King, C. A. M., 1972, Beaches and coasts: Arnold Publish. Ltd., London.

McGill, J. T., 1958, Map of coastal landforms of the world. – Geogr. Rev., 48: 402–405.
(1959): Coastal classification maps: a review. – 2nd Coastal Geogr. Conference: 1–21.
(1960a): Map in W. C. Putnam et al.: Natural coastal environments of the world. – Los Angeles.
(1960b): Selected bibliography of coastal geomorphology of the world. – Los Angeles.

Murray, G. L., 1961, Geology of the Atlantic and Gulf Coast province of North America: Harpers Geosci. Ser., New York, 692 p. (Bibliogr. 2000 titles).

Nelson, B. W., (ed), 1972, Environmental framework of coastal plain estuaries, based on presentations at the symposium on estuaries, 18th annual meeting of the southeastern section The Geological Society of America, Columbia, South Carolina, April 10 and 11, 1969: *Geol. Soc. America Mem.* 133.

Putnam, W. C., 1960, Natural coastal environments of the world: Univers. Calif., Los Angeles.

Shepard, F. P. and H. R. Wanless, 1971, Our changing coastlines: McGraw-Hill, New York, 579 p.

Verger, F., 1968, Marais et wadden du littoral français, étude de géomorphologie littorale: Biscaye frères, imprimeurs, Bordeaux, 544 p., 2 maps.

Wüst, G., 1964, Stratification and circulation in the Antillen-Caribbean basins, part I, Spreading and mixing of the water types, with an oceanographic atlas: New York, 201 p.

Zenkovich, V. P., 1967, Processes of coastal development: Oliver & Boyd, London.

Tidal current tables, 1961: US Department of Commerce, Coastal and Geodetic Service, US Government Printing, Washington D. C.

Coasts, lagoons and estuaries

Allen, J. R. L., 1968, Current ripples: North Holland Publish. Comp.

Carlson, P. R., T. J. Conomos, R. J. Janda and D. H. Peterson, 1975, Principal sources and dispersal patterns of suspended particulate matter in nearshore surface waters of the Northeast Pacific Ocean: NASA-CR-142641, Geol. Survey, Menlo Park, Calif.

Emery, K. O., R. E. Stevenson and Hedgpeth, 1957, Estuaries and lagoons, treatise on marine ecology and paleoecology, vol. I Ecology: *Geol. Soc. America Mem.* 67, 673 p.

Gierloff-Emden, H. G., 1959, Die Küste von El Salvador. Eine morphologisch-ozeanographische Monographie: *Acta Humboldtiana* no. 2, Wiesbaden, 183 p.

Gierloff-Emden, H. G., 1959, Lagunen, Nehrungen, Strandwälle und Flußmündungen im Geschehen tropischer Flachlandküsten: *Zeitschr. Geomorph.*, vol. 3, I, p. 29–46.

Ginsburg, R. N. (ed.), 1974, Tidal deposits, a casebook of recent examples and fossil counterparts: Springer Publish., New York.

Lauff, G. H. (ed.), 1967, Estuaries: *Am. Assoc. Advancem. Sci.*, Publ. no. 83, Washington.

Perkins, E. J., 1974, The biology of estuaries and coastal waters: Academic Press, London and New York, 640 p.

Reineck, H. E. and I. B. Singh, 1973, Depositional sedimentary environments: Springer Verlag, Berlin–Heidelberg, 439 p., 579 fig.

Schäfer, W., 1972, Ecology and palaeoecology of marine environments: Oliver & Boyd, Edinburgh, 568 p.

Swift, D. J. P., P. B. Duane and D. H. Pilkey (ed.), 1972, Shelf sediment transport, process and pattern: Dowden, Hutchinson & Ross Inc., Pennsylvania.

Tait, R. V., 1962, Elements of marine ecology: Plenum Press, New York, 272 p.

Coastal resources

Detwyler, Th., 1971, Man's impact on environment: McGraw Hill, New York, 731 p.

Gunter, G., B. S. Ballard and A. Venkataramiah, 1974, A review of salinity problems of organisms in United States coastal areas subject to the effects of engineering works: *Gulf Research Reports*, vol. 4, no. 3, pp. 380–475.

Inman, 1973, The coastal challenge: *Science*, vol. 181, no. 4094, p. 20–32.

Mitchell, J. K., 1974, Community response to coastal erosion, individual and collective adjustment to hazard on the Atlantic shore: Univers. of Chicago, Dep. of Geogr., Res. Pap. no. 156, 209 p.

Ruivo, M. (ed.), 1972, Marine pollution and sea life: FAO of UNO, Fishing News (Books) Ltd., London.

Walker, H. J. (ed.), 1975, Coastal resources: *Geoscience and man*, vol. XII, School of Geoscience, Louisiana State Univers., Baton Rouge, pp. 115.

Mapping with remote sensing

Colvocoresses, A. P., 1972, Earth resources observation systems, cartographic applications program: Annual report to the National Aeronautics and Space Administration, FY 1972, Topographic Division, Geol. Surv., US Dep. of the Interior, Washington D. C.

Doyle, F. J., 1972, Can satellite photography contribute to topographic mapping? : *US Geol. Surv. Journ. Res.*, vol. 1, no. 2.

Gettys, R. F., 1966, Extraction of nautical chart information from color photographs obtained on Gemini orbital flights 4, 5 and 7: Proc. Fourth Sympos. Remote Sensing of Environm., Michigan.

Gierloff-Emden, H. G. and U. Rust, 1971, Verwertbarkeit von Satellitenbildern für geomorphologische Kartierungen in Trockenräumen, Chihuahua, New Mexico, Baja California, Bildinformationen und Geländetest: *Münchener Geogr. Abhdl.*, vol. 5.

Hake, G., Der Informationsgehalt der Karte – Merkmal und Maße: in E. Arnberger (ed.), Grundsatzfragen der Kartographie.

d'Hollander, R., 1960, Problemes de photogeographie et de representation cartographique du littoral Atlantique Français: XIXe Congres Intern. de Geogr. Stockholm – août 1960, Institut Geographique National, Paris.

Koeman, C., 1970, The impact of photography from space on small scale and atlas cartography: in Intern. Jahrb. f. Kartographie, p. 35–40.

Louis, H., 1958, Die Karte als wissenschaftliche Ausdrucksform: Tag.-Ber. Deutsch. Geogr.-Tag 1957, Wiesbaden.

Lyon, R. J. P., 1970, Multiband approach to geological mapping from orbiting satellites: in Remote sensing of environment, vol. I, p. 237–244.

Mott, P. G., H. Fullard, J. C. Bartholomew, J. D. Leatherdale and H. J. Chisom, 1975, Study of techniques and applications of satellite imagery to small scale mapping: Hunting Surveys, Boreham Wood (England), NASA CR-142640, 65 p.

Poleyn, F. C. and D. R. Lyzenga, 1973, Updating coastal and navigational charts using ERTS–1 data: Third ERTS Sympos. Abstr., NASA, Goddard Space Flight Center, Greenbelt, Maryland.

Schmidt-Falkenberg, H., 1970, Interne und integrale Photointerpretation: *B. u. L.* 38, p. 313–318.

Schmidt-Falkenberg, H., 1972, Experimentelle Untersuchungen in der Photointerpretation: Bericht XII, Intern. Kongress Ottawa, 13 p.

Wobber, F. J., 1972, The use of orbital photography for earth resources satellite mission planning: *Photogrammetria*, vol. 28, p. 35–59.

Bei Fragen zur Produktsicherheit wenden Sie sich bitte an:
If you have any questions regarding product safety,
please contact:

Walter de Gruyter GmbH
Genthiner Straße 13
10785 Berlin
productsafety@degruyterbrill.com